An Introduction to Chemical Kinetics

I0047713

An Introduction to Chemical Kinetics

Claire Vallance
Department of Chemistry, University of Oxford

Morgan & Claypool Publishers

Copyright © 2017 Morgan & Claypool Publishers

All rights reserved. No part of this publication may be reproduced, stored in a retrieval system or transmitted in any form or by any means, electronic, mechanical, photocopying, recording or otherwise, without the prior permission of the publisher, or as expressly permitted by law or under terms agreed with the appropriate rights organization. Multiple copying is permitted in accordance with the terms of licences issued by the Copyright Licensing Agency, the Copyright Clearance Centre and other reproduction rights organisations.

Rights & Permissions
To obtain permission to re-use copyrighted material from Morgan & Claypool Publishers, please contact info@morganclaypool.com.

ISBN   978-1-6817-4664-7 (ebook)
ISBN   978-1-6817-4667-8 (print)
ISBN   978-1-6817-4666-1 (mobi)

DOI   10.1088/978-1-6817-4664-7

Version: 20170901

IOP Concise Physics
ISSN 2053-2571 (online)
ISSN 2054-7307 (print)

A Morgan & Claypool publication as part of IOP Concise Physics
Published by Morgan & Claypool Publishers, 1210 Fifth Avenue, Suite 250, San Rafael, CA, 94901, USA

IOP Publishing, Temple Circus, Temple Way, Bristol BS1 6HG, UK

To the many Oxford MChem students who have directly or indirectly shaped the course on which this book is based.

Contents

Preface

An Introduction to Chemical Kinetics began life as a set of lecture notes for a core first year lecture course at the University of Oxford. In turning the notes into a book, I have reordered and extended some of the material considerably. This has been an interesting and useful exercise for me, and possibly for future Oxford chemists (I shall let them be the judge), as next year's lecture course will certainly be modified based on the experience of writing this book.

The book is intended to give a relatively concise introduction to the main principles of Chemical Kinetics at a level suitable for any undergraduate chemistry student. Elementary reactions and their rates are introduced and explained using simple collision theory. Rate laws and reaction mechanisms are covered in some detail, as are experimental methods and data analysis techniques for obtaining kinetic information across a broad range of timescales. More advanced topics are touched on very briefly at various points in the text, but for the most part are left for the reader to explore in more comprehensive texts.

Claire Vallance, July 2017

Acknowledgments

I would like to thank the many students I have lectured to in Oxford's Department of Chemistry and tutored at Hertford College. Their questions and comments over the years have indirectly shaped a number of aspects of this book. I would also like to thank Nicki Dennis, of Morgan and Claypool Publishing, for convincing me that anyone other than Oxford Chemistry undergraduates would be interested in reading my take on chemical reaction rates.

Author biography

Claire Vallance

Claire Vallance is a Professor of Physical Chemistry in the Department of Chemistry at the University of Oxford, and Tutorial Fellow in Physical Chemistry at Hertford College, Oxford. She was brought up in the UK and New Zealand, and holds BSc (hons) and PhD degrees from the University of Canterbury (Christchurch, NZ), where she worked on gas-phase molecular dynamics. Her current research interests include reaction dynamics, the use of optical microcavities in chemical sensing applications, and the development of spectroscopic techniques for use during cardiovascular surgery and neurosurgery. She has given lecture courses on chemical kinetics, properties of gases, symmetry and group theory, reaction dynamics, and astrochemistry, as well as numerous outreach and public engagement lectures, and her tutorial teaching spans the breadth of physical chemistry. She is the author of over 90 journal articles, four book chapters, nine patents, an e-Textbook on *Symmetry and Group Theory*, and a textbook *Astrochemistry: from the Big Bang to the Present Day*, and also co-edited the textbook *Tutorials in Molecular Reaction Dynamics*.

An Introduction to Chemical Kinetics

Claire Vallance

Chapter 1

Elementary reactions

1.1 Introduction

Chemical reactions occur over a vast range of different timescales. At one extreme, the chemical reactions involved in fossilisation occur over thousands of years. In contrast, many of the reactions that are important in combustion, atmospheric chemistry, and biology occur on a timescale of a few hundred femtoseconds (1 fs $= 10^{-15}$ s). One of the aims of any book on chemical reaction rates must therefore be to explain the enormous range of chemical timescales. In addition, we will find that experimental and theoretical studies into the rates of chemical reactions can also provide a great deal of insight into the detailed mechanisms by which chemical reactions proceed.

To begin, we must define what we mean by the 'rate' of a chemical reaction. In simple terms, the reaction rate can be defined as the rate at which the concentrations of the reactants or products change with time. This is a relatively straightforward property to measure experimentally. If we follow the concentration of a reactant or a product as a function of time and plot the data as shown in figure 1.1, the reaction rate at any given time is simply the slope of the graph. Note that the slope, and therefore the reaction rate, is not constant with time. The rate decreases as the reactants are used up, implying that the rate depends on the concentrations of the reactants. Determining this concentration dependence turns out to be key to gaining insight into the reaction mechanism, and will be treated in detail in later chapters. We can gain some initial insight into both the concentration dependence of the rate and the widely varying rates for different reactions by considering the factors that might influence the rate of a simple bimolecular reaction, in which two reactants collide and react to form products. The rate at which reaction occurs in this case is determined both by the number of collisions per second, or *collision frequency* (i.e. the number of 'chances' for the reaction to occur per second), and by the fraction of those collisions that lead to successful reaction. The collision frequency is governed mostly by the reactant concentrations, while the reaction probability is

doi:10.1088/978-1-6817-4664-7ch1   1-1   © Morgan & Claypool Publishers 2017

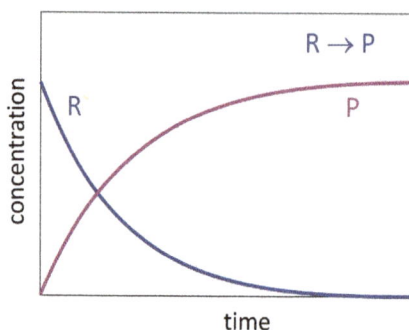

Figure 1.1. Reactant and product concentrations as a function of time for a simple reaction R $\longrightarrow$ P.

determined primarily by the presence of any energetic barriers along the reaction pathway and the amount of energy available for the reactants to overcome them.

In the remainder of this chapter, we will investigate the simplest types of chemical reaction, known as *elementary reactions*, and will develop a simple model that allows us to predict reaction rates, and to explain, at least to first order, the variety of timescales over which chemical reactions are observed to occur.

In chapter 2, we will introduce rate laws and rate constants, which relate the rate of a chemical reaction to the concentrations of the chemical species present and provide the mathematical framework for understanding chemical reaction rates.

In chapter 3, we will introduce a number of methods for determining the rate law for a reaction from data recorded during experimental measurements of the reaction rate. We will also explore how data on the temperature dependence of the rate constant can provide information on activation barriers along the reaction pathway. Experimental methods for following reactant and product concentrations over a broad range of timescales will be introduced in chapter 4.

As noted previously, kinetics studies provide a powerful tool for elucidating the sequence of chemical steps involved in a reaction. After introducing multi-step reaction mechanisms in chapter 5, we will look at numerous mechanistic examples of increasing complexity, in chapters 6 and 7.

1.2 Elementary reactions

An elementary reaction is the simplest kind of reaction, and occurs in a single step. Elementary reactions are generally classified in terms of the number of reactant molecules involved, known as the *molecularity* of the reaction. A *unimolecular* reaction involves a single reactant; for example, the *cis* to *trans* isomerisation of but-2-ene.

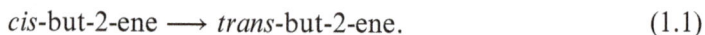

$$cis\text{-but-2-ene} \longrightarrow trans\text{-but-2-ene}. \tag{1.1}$$

Dissociation reactions also fall into this category. For example

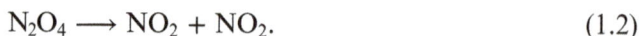

$$N_2O_4 \longrightarrow NO_2 + NO_2. \tag{1.2}$$

A *bimolecular* reaction involves a reactive collision between two chemical species, and represents the most common type of elementary reaction. For example:

$$F^- + CH_3Cl \longrightarrow CH_3F + Cl^-. \tag{1.3}$$

Very rarely, under conditions of extremely high pressure, a termolecular step may occur, which involves simultaneous collision of three reactant molecules. We will see examples of termolecular steps in reaction mechanisms in sections 7.2.3 and 7.4.1. In general, steps in a mechanism that appear to be termolecular should be treated with some suspicion, as more often than not they are actually the result of two sequential bimolecular steps. This is explored in more detail in section 6.4.

Any chemical process may be broken down into a sequence of one or more elementary reactions, also sometimes known as elementary processes or elementary steps. A reaction involving more than one elementary step is known as a *complex reaction*. It is important to note that most chemical reactions you will be familiar with are in fact complex reactions. The overall reaction equations for such processes are the net result of the elementary steps involved in the mechanism, and there is often no single chemical process that corresponds to the reaction equation. For example, many readers will be familiar with the well-known Haber–Bosch process for producing ammonia, NH_3 from molecular nitrogen and hydrogen. The overall reaction equation is:

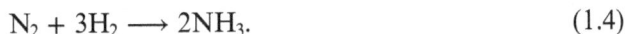

$$N_2 + 3H_2 \longrightarrow 2NH_3. \tag{1.4}$$

However, the reaction actually proceeds via a large number of elementary steps, involving a number of different atomic and molecular species bound to, or 'adsorbed onto', particles of the iron-based catalyst. These bound species are denoted by the subscript, (ads), in the mechanism below, with gas-phase species denoted by the subscript, (g).

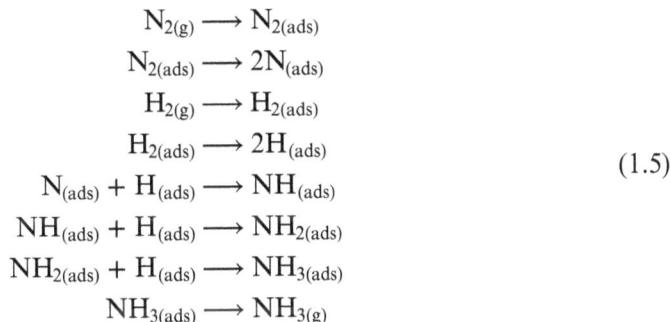

$$
\begin{aligned}
N_{2(g)} &\longrightarrow N_{2(ads)} \\
N_{2(ads)} &\longrightarrow 2N_{(ads)} \\
H_{2(g)} &\longrightarrow H_{2(ads)} \\
H_{2(ads)} &\longrightarrow 2H_{(ads)} \\
N_{(ads)} + H_{(ads)} &\longrightarrow NH_{(ads)} \\
NH_{(ads)} + H_{(ads)} &\longrightarrow NH_{2(ads)} \\
NH_{2(ads)} + H_{(ads)} &\longrightarrow NH_{3(ads)} \\
NH_{3(ads)} &\longrightarrow NH_{3(g)}
\end{aligned}
\tag{1.5}
$$

We see that there is never actually a reaction between a molecule of N_2 and a molecule of H_2, despite this apparently being implied by the overall reaction equation. In later chapters, we will look at many examples of multi-step mechanisms, and ways in which we can relate the sequence of steps to the measured rate of the overall reaction.

1.3 The rates of elementary reactions: energetic considerations

Most chemical reactions have an energetic barrier known as the *activation barrier* somewhere along the reaction pathway, as shown schematically in figure 1.2. Even if

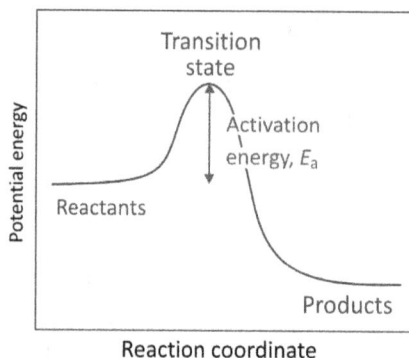

Figure 1.2. Schematic of a reaction potential energy pathway for an exothermic reaction, showing the energies of the reactants, transition state, and products. The difference in energy E_a between the transition state and reactants is known as the *activation barrier* or *activation energy*.

a reaction is highly exothermic, meaning that products are thermodynamically, vastly favoured over reactants, there can still be significant energetic barriers to product formation. For example, a reaction is often initiated by cleavage of a particular chemical bond, a process that requires energy in order to occur. Some reactions, such as radical–radical or ion–molecule reactions, have essentially no activation barrier for the reaction itself. However, there may still be small energetic barriers associated with other processes involved in the reaction. These include barriers associated with the reactants diffusing together through the gas-phase or through solution, as well as the more exotic 'centrifugal barriers' associated with the conservation of angular momentum arising from the orbital motion of the reactants in the gas phase. In light of these considerations, in energetic terms, we can think of an elementary reaction as a transition between two atomic or molecular states separated by an energetic barrier. The barrier is variously referred to as a *potential barrier*, *activation barrier*, or *activation energy*, and determines the rate at which the reaction will occur at a particular temperature. When the barrier is low, the thermal energy of the reactants will generally be high enough to surmount the barrier and move over to the products, and the reaction will be fast. However, when the barrier is high, only a small fraction of reactants will have sufficient energy, and the reaction will be much slower. If we increase the temperature of the reaction mixture, a greater fraction of reactants will have sufficient energy to surmount the barrier and react, and the reaction rate will increase. In the following section, we will consider a simple model that allows us to predict quantitatively the effect of temperature on the reaction rate.

1.4 The rates of elementary reactions: simple collision theory and the Arrhenius equation

Simple collision theory, as the name suggests, is the simplest model that has been developed in order to understand and predict the reaction rate for an elementary bimolecular reaction $A + B \longrightarrow P$. Much more sophisticated models are available, which provide more accurate rate predictions, but simple collision theory contains the basic elements needed to understand the timescales over which chemistry occurs and to provide some insight into the temperature dependence of chemical reaction rates.

We begin by considering the factors we might expect a reaction rate to depend upon. Obviously, the rate of reaction must depend upon the rate of collisions between the reactants. However, not every collision leads to a reaction. Some colliding pairs do not have enough energy to overcome the activation barrier, and any theory of reaction rates must take this energy requirement into account. Also, it is highly likely that a reaction will not even take place on every collision for which the energy requirement is met. The reactants may need to collide in a particular orientation, for example, or some of the energy may need to be present in a particular form, perhaps as vibrational excitation in a bond coupled to the reaction coordinate. In summary, there are three factors that influence whether or not a successful reactive collision will occur, and we might expect an expression for the rate of a bimolecular reaction to take the following form:

$$\text{rate} = (\text{encounter rate}) \, (\text{energy requirement}) \, (\text{steric requirement}). \quad (1.6)$$

We will now consider each of these factors in more detail.

1.4.1 Encounter rate or collision rate

The rate of collisions between two potential reactants, i.e. the number of collisions per second, depends on three factors:

1. The number of reactant molecules per unit volume, otherwise known as the number density. This is the gas-phase equivalent of concentration. The more particles per unit volume, the more often collisions will occur.

2. The sizes of the two collision partners—the larger the particles in a sample of gas, the more often they will collide with each other. This is encapsulated in a quantity known as the collision cross section, σ_C, which has units of area. The collision cross-section can be thought of as the 'target area' that one collision partner presents to the second, as shown in figure 1.3. For a collision between two 'hard spheres' A and B, the radius of the target area is simply the sum of the radii r_A and r_B of the two particles; a quantity known as the *collision diameter*, $d = r_A + r_B$, with $\sigma_C = \pi d^2$.

3. The relative velocity of the collision partners. The faster the particles are moving around in a gas, the more often they collide. For gas molecules at thermal equilibrium, the distribution of the relative velocities $P(v_{\text{rel}})$ is given by the Maxwell–Boltzmann distribution,

$$P(v_{\text{rel}})dv_{\text{rel}} = 4\pi \left(\frac{\mu}{2\pi k_B T} \right)^{3/2} v_{\text{rel}}^2 \exp\left(-\frac{\mu v_{\text{rel}}^2}{2k_B T} \right) dv_{\text{rel}} \quad (1.7)$$

where $\mu = m_A m_B / (m_A + m_B)$ is the reduced mass of the two collision partners, k_B is Boltzmann's constant, and T is the temperature.

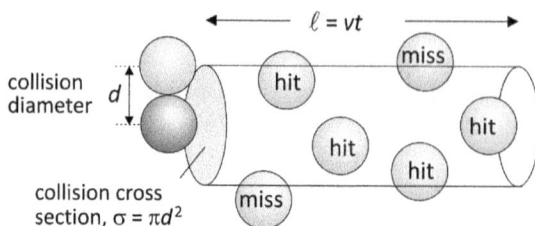

Figure 1.3. The collision cross section σ_C for an A + B collision is the effective cross-sectional 'target' area around A that B must enter in order for the two particles to collide. For a hard-sphere collision, this is simply a circle with diameter d equal to the sum of the radii of the two colliding particles. In a time t, a particle travelling in a straight line with velocity v will travel a distance $l = vt$, and will collide with every particle whose centre lies within the cylindrical volume $\sigma_C l$. To obtain accurate collision frequencies for particles in a gas, we replace v with the mean relative velocity, $\langle v_{rel} \rangle$ of the gas particles.

Taking the average of this distribution, we find that the mean relative velocity $\langle v_{rel} \rangle$ for molecules with a Maxwell–Boltzmann distribution of velocities is

$$\langle v_{rel} \rangle = \left(\frac{8k_B T}{\pi \mu} \right)^{1/2}. \tag{1.8}$$

Referring again to figure 1.3, we see that, in a time t, the particle of interest will travel on average a distance of $d = \langle v_{rel} \rangle t$, and will therefore collide with every particle within the cylindrical volume $V = \sigma_C d = \sigma_C \langle v_{rel} \rangle t$. To determine the collision frequency, or number of collisions per second, all we need to do is to determine the number of particles within the volume corresponding to a time of $t = 1$ s. Since we know the number densities of the particles, we simply multiply the cylindrical volume of interest by the number of particles per unit volume in order to determine the number of collisions, i.e. the collision frequency Z_{AB} is simply the product of the three factors considered above:

$$Z_{AB} = \sigma_C \left(\frac{8k_B T}{\pi \mu} \right)^{1/2} n_A n_B. \tag{1.9}$$

Note that if we are considering collisions between two particles of the same type, i.e. A + A rather than A + B, we need to multiply the expression by a factor of 1/2 to avoid counting each collision twice.

$$Z_{AA} = \frac{1}{2} \sigma_C \left(\frac{8k_B T}{\pi \mu} \right)^{1/2} n_A^2. \tag{1.10}$$

1.4.2 Energy requirement

From the Maxwell–Boltzmann distribution of relative molecular speeds, we can also work out the fraction of collisions for which the collision energy is high enough to overcome any activation barrier, E_a for the reaction. This turns out to be given simply by the factor $\exp(-E_a/k_B T)$.

1.4.3 Steric requirement

Experimentally, measured rates are often found to be up to an order of magnitude smaller than those calculated from simple collision theory, suggesting that features, such as the relative orientation of the colliding species, are important in determining the reaction rate. We account for the disagreement between the experiment and theory by introducing a steric factor, P, into our expression for the reaction rate. Alternatively, we can replace the collision cross section, σ_C, with a reaction cross section σ_R, where $\sigma_R = P\sigma_C$. Usually, P is considerably less than unity, but values greater than one are also possible. An example is found in the 'harpoon reaction' between Rb and Cl_2. The reaction mechanism involves an electron transfer at large separations to form $Rb^+ + Cl_2^-$, after which the electrostatic attraction between the two ions guarantees a reaction. P is large because the reaction cross section is determined by the electron transfer distance, which is much larger than the collision diameter.

1.4.4 Putting everything together

Combining these three terms, the simple collision theory expression for the reaction rate is:

$$\text{rate} = P\sigma_C \left(\frac{8k_B T}{\pi\mu}\right)^{1/2} \exp\left(-\frac{E_a}{RT}\right) n_A n_B. \tag{1.11}$$

Unsurprisingly, we find that the reaction rate is proportional to the concentrations (or number densities) n_A and n_B of the two reactants. The temperature-dependent factor that relates the rate to the concentrations is known as the rate constant, usually given by the symbol k or $k(T)$.

$$k(T) = P\sigma_C \left(\frac{8k_B T}{\pi\mu}\right)^{1/2} \exp\left(-\frac{E_a}{RT}\right). \tag{1.12}$$

It is very important to note that *the rate constant is only a constant at a fixed temperature.*

The rate constant for an elementary reaction determines how quickly or slowly the reaction will proceed relative to other reactions. The dominant factor in determining the rate constant for a given process is usually the energetic factor, which in turn is determined by the activation energy, E_a. Rate constants for elementary processes will become very important from the next chapter onwards, when we introduce the idea of a rate law.

1.4.5 Catalysis

A catalyst is a substance that is added to a reaction mixture in order to speed up the reaction. The catalyst provides an alternative reaction mechanism with a lower activation barrier, but is regenerated as part of the reaction mechanism, allowing it to catalyse a reaction many times over. One mechanism for catalysis, involving binding of reactant molecules to an enzyme, is considered later in the book in section 6.5.

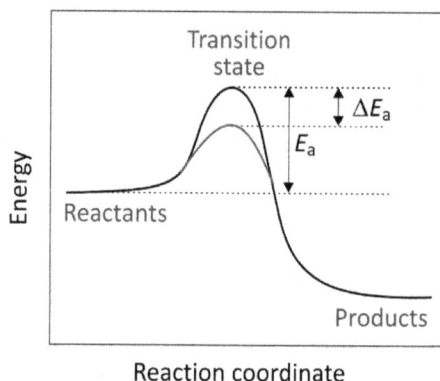

Figure 1.4. A catalyst lowers the activation barrier E_a by an amount ΔE_a.

In the previous section, we determined that the rate constant has an exponential dependence on the activation energy. This allows us to make quantitative predictions about the effect of a catalyst, assuming we have some knowledge of the extent to which the activation energy is lowered. Consider a reaction with activation energy E_a, which is lowered by an amount ΔE_a on addition of a catalyst (see figure 1.4 for a potential energy diagram). For the uncatalysed reaction, the rate constant follows the proportionality

$$k_{\text{uncat}}(T) \propto \exp\left(-\frac{E_a}{RT}\right) \tag{1.13}$$

while for the catalysed reaction, we have

$$k_{\text{cat}}(T) \propto \exp\left(-\frac{E_a - \Delta E_a}{RT}\right). \tag{1.14}$$

To determine the factor by which the reaction is sped up on addition of the catalyst, we take the ratio of the two rate constants.

$$\frac{k_{\text{cat}}}{k_{\text{uncat}}} = \frac{\exp\left(-\frac{(E_a - \Delta E_a)}{RT}\right)}{\exp\left(-\frac{E_a}{RT}\right)} = \exp\left(\frac{\Delta E_a}{RT}\right). \tag{1.15}$$

Addition of the catalyst therefore increases the reaction rate by a factor of $\exp(\Delta E_a / RT)$

1.4.6 Shortcomings of simple collision theory

Though simple collision theory can explain many features of chemical reaction rates, including the effect of adding a catalyst to a reaction mixture, it generally does not predict absolute rate constants in quantitative agreement with experiment.

Experimentally, the temperature dependence of the rate constant is often described well by an empirical expression known as the *Arrhenius equation*

$$k(T) = A \exp\left(-\frac{E_a}{RT}\right) \qquad\qquad (1.16)$$

where A is known as the *pre-exponential factor*. We will return to the Arrhenius equation in more detail later in section 3.6.

Comparing equation 1.16 with equation 1.12, we see that simple collision theory is able to rationalise the functional form of the Arrhenius equation; an important achievement. However, at a quantitative level, the predictions of the theory are often far from accurate. There are a number of ways in which the model breaks down.

1. It does not account for the fact that, unless the collision is head on, not all of the kinetic energy of the two reactants is available for reaction. Conservation of linear and angular momentum means that only the kinetic energy corresponding to the velocity component along the relative velocity vector of the reactants actually contributes to the collision energy.

2. The energy stored in the internal degrees of freedom in the reactants (vibrations, rotations, etc) has been ignored. For reactions involving large molecules, this often leads to a sizeable discrepancy between simple collision theory and experiment, though this is partly corrected for by the inclusion of the steric factor, P. This shortcoming is addressed in another model, known as transition-state theory, which is based on principles of statistical mechanics. Transition-state theory is beyond the scope of this book, but is commonly taught in advanced undergraduate chemistry courses, and is described in many textbooks on statistical mechanics.

Now that we have considered the various factors influencing the rate of an elementary reaction, in the next chapter we will introduce the general framework of rate laws and reaction orders used to describe and quantify the kinetics of both elementary or multi-step reactions.

Chapter 2

Rate laws: relating the reaction rate to reactant concentrations

In the previous chapter, we considered the factors that determine the rate of an elementary bimolecular reaction, and showed that the rate was proportional to the concentrations (or number densities) of the reactants. In the present chapter, we will formalise this relationship by introducing the concept of a rate law, which relates the reaction rate to the concentrations of the various species present in the reaction mixture. Rate laws will form the basis of a framework within which we can describe the rates of both elementary and complex (multi-step) reactions. We will begin with a formal definition of the rate of reaction.

2.1 Rate of reaction

As noted earlier, the rate of a chemical reaction is defined as the rate at which reactants are used up, or equivalently, the rate at which products are formed. The rate therefore has units of concentration per unit time, mol dm^{-3} s^{-1}. For gas-phase reactions, alternative units of 'concentration' are often used, usually units of pressure, such as Torr, mbar, or Pa. The resulting rates then have dimensions of Torr s^{-1}, mbar s^{-1}, or Pa s^{-1}. Conversion between a rate expressed in units of pressure and one in units of concentration is straightforward: the ideal gas law, $pV = nRT$, can be rearranged to give $n/V = p/RT$. We therefore find that a rate in pressure units can be converted to a rate in concentration units simply by dividing by RT.

Given the above definition of the rate of reaction, to measure a reaction rate, we simply need to monitor the concentration of any one of the reactants or products as a function of time. This raises a potential problem with our definition, which we will explore by means of an example. Consider again the well-known Haber–Bosch process, used industrially to produce ammonia:

doi:10.1088/978-1-6817-4664-7ch2   2-1   © Morgan & Claypool Publishers 2017

$$N_2 + 3H_2 \longrightarrow 2NH_3. \tag{2.1}$$

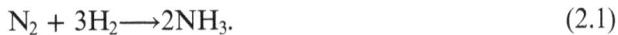

The reactant N_2 has a stoichiometric coefficient of 1, H_2 has a coefficient of 3, and NH_3 has a coefficient of 2. Experimentally, we could determine the rate of this reaction in any one of three ways, by monitoring the changing concentration of N_2, H_2, or NH_3. Say we monitor the concentration of N_2, and obtain a rate of $-d[N_2]/dt = x$ mol dm^{-3} s^{-1}. Since for every mole of N_2 that reacts, we lose three moles of H_2, if we had monitored H_2 instead of N_2, we would have obtained a rate of $-d[H_2]/dt = 3x$ mol dm^{-3} s^{-1}. Similarly, monitoring the concentration of NH_3 would yield a rate of $2x$ mol dm^{-3} s^{-1}. Clearly, the same reaction cannot have three different rates, so we appear to have a problem. The solution is actually very simple: the reaction rate is defined as the rate of change of the concentration of a reactant or product divided by its stochiometric coefficient. For the above reaction, the rate is therefore

$$\text{rate} = -\frac{d[N_2]}{dt} = -\frac{1}{3}\frac{d[H_2]}{dt} = \frac{1}{2}\frac{d[NH_3]}{dt}. \tag{2.2}$$

Note that a negative sign appears when we define the rate using the concentration of one of the reactants. This is because the rate of change of a reactant is negative (since it is being used up in the reaction), but the reaction rate needs to be a positive quantity.

2.2 Rate laws

We have discovered already that the reaction rate depends on the concentrations of the chemical species present in the reaction mixture. This relationship is quantified in the rate law for the reaction. Chemical species appearing in a rate law may include reactants, products, and catalysts, but not reactive intermediates. Reactive intermediates are species that appear in elementary steps of the mechanism, but not in the overall reaction equation. We will look at these in some detail later in chapters 5 to 7.

Many reactions follow a simple rate law, which takes the form

$$\text{rate} = k[A]^a[B]^b[C]^c... \tag{2.3}$$

i.e. the rate is proportional to the product of the concentrations of the reactants, each raised to some power, with the rate constant k providing the constant of proportionality. The power a particular concentration is raised to is the *order* of the reaction with respect to that reactant. The reaction described by the rate law in equation (2.3) has order a with respect to reactant A, order b with respect to reactant B, and so on. Note that when we are dealing with multi-step reactions, the orders do not have to be integers. The sum of the individual orders is called the *overall order*.

Elementary reactions always follow simple rate laws, with the overall order reflecting the molecularity of the process. We can write down the rate law for an elementary reaction directly from the reaction equation, since the rate is simply proportional to the concentrations of each reactant. For example:

Unimolecular reaction $\quad$ A $\longrightarrow$ B $\qquad$ rate $= -\dfrac{d[A]}{dt} = k[A]$

Bimolecular reaction $\qquad$ A + B $\longrightarrow$ P $\qquad$ rate $= -\dfrac{d[A]}{dt} = k[A][B]$

$\qquad\qquad\qquad\qquad$ A + A $\longrightarrow$ P $\qquad$ rate $= -\dfrac{d[A]}{dt} = k[A][A] = k[A]^2.$

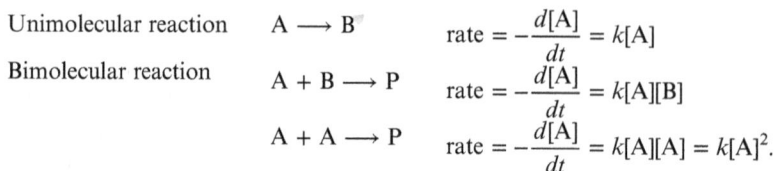

A unimolecular elementary reaction therefore has an overall order of one, while a bimolecular reaction has an order of one with respect to each reactant, and an overall order of two.

Many complex (multi-step) reactions also follow simple rate laws, though, in these cases, the orders will not necessarily reflect the stoichiometry of the reaction equation. For example,

$$H_2 + I_2 \longrightarrow 2HI \qquad \text{rate} = k[H_2][I_2]$$
$$3ClO^- \longrightarrow ClO_3^- + 2Cl^- \qquad \text{rate} = k[ClO^-]^2.$$

Other complex reactions follow complex rate laws. These often have a much more complicated dependence on the chemical species present, and may also contain more than one rate constant. While simple rate laws may imply either a simple (single-step) or complex (multi-step) mechanism, complex rate laws always imply a multi-step reaction mechanism. An example of a reaction with a complex rate law is

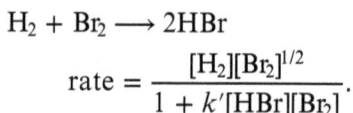

$$H_2 + Br_2 \longrightarrow 2HBr$$
$$\text{rate} = \frac{[H_2][Br_2]^{1/2}}{1 + k'[HBr][Br_2]}.$$

Inspection of the rate law reveals that the reaction has order 1 with respect to $[H_2]$, but that it is impossible to define orders with respect to Br_2 and HBr, since there is no direct proportionality between their concentrations and the reaction rate. Consequently, it is also impossible to define an overall order for this reaction. To give the reader some idea of the complexity that may underlie an overall reaction equation, a slightly simplified version of the sequence of elementary steps involved in the above reaction is shown below. We will return to this reaction later when we look at chain reactions in chapter 7.

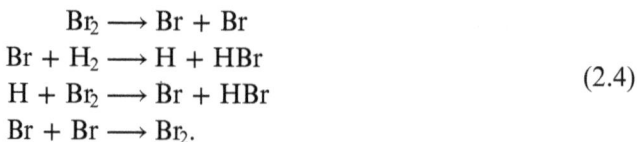

$$
\begin{aligned}
Br_2 &\longrightarrow Br + Br \\
Br + H_2 &\longrightarrow H + HBr \\
H + Br_2 &\longrightarrow Br + HBr \\
Br + Br &\longrightarrow Br_2.
\end{aligned}
\qquad (2.4)
$$

To summarise, elementary reactions follow simple rate laws, while multi-step processes may follow simple or complex rate laws. As the above examples have hopefully illustrated, in these cases the rate law generally does not follow from the overall reaction equation. This should not come as a surprise: as noted earlier, the overall reaction equation for a multi-step process is simply the net result of all of the elementary reactions in the mechanism, and the reactants appearing in the overall reaction equation may not actually react directly with each other in any of the elementary steps.

Even though the rate law for a multi-step reaction cannot immediately be written down from the reaction equation as it can in the case of an elementary reaction, the rate law *is* a direct result of the sequence of elementary steps that constitute the reaction mechanism. As such, it provides one of our best tools for determining an unknown mechanism. As we will find out later, once we know the sequence of elementary steps that constitute the reaction mechanism, we can quite quickly deduce the rate law. Conversely, if we do not know the reaction mechanism, we can carry out experiments to determine the orders with respect to each reactant (see chapters 3 and 4) and then try out various 'trial' reaction mechanisms to see which one fits best with the experimental data.

2.3 The units of the rate constant

A point that often seems to cause considerable confusion when dealing with rate laws and rate constants is the fact that the units of a rate constant depend on the form of the particular rate law in which it appears, i.e. a rate constant appearing in a first order rate law will have different units from a rate constant appearing in a second order or third order rate law. This follows immediately from the fact that the reaction rate always has the same units of concentration per unit time, which must match the overall units of a rate law in which concentrations raised to varying powers may appear. The good news is that it is very straightforward to determine the units of a rate constant in any given rate law. We will illustrate this with a few examples. Consider the second order rate law

$$\text{rate} = k[\text{H}_2][\text{I}_2]. \tag{2.5}$$

If we substitute the appropriate units into the equation for the rate and the concentrations, and use $\{k\}$ to denote the units of the rate constant, we obtain

$$(\text{mol dm}^{-3}\,\text{s}^{-1}) = \{k\}(\text{mol dm}^{-3})(\text{mol dm}^{-3}). \tag{2.6}$$

Rearranging this expression yields the units of the rate constant.

$$\{k\} = \frac{(\text{mol dm}^{-3}\,\text{s}^{-1})}{(\text{mol dm}^{-3})(\text{mol dm}^{-3})} = \text{mol}^{-1}\,\text{dm}^3\,\text{s}^{-1}. \tag{2.7}$$

We can apply the same treatment to a first order rate law, for example

$$\text{rate} = k[\text{CH}_3\text{N}_2\text{CH}_3]. \tag{2.8}$$

Substituting in the relevant units gives

$$(\text{mol dm}^{-3}\,\text{s}^{-1}) = \{k\}(\text{mol dm}^{-3}) \tag{2.9}$$

which we can solve to give

$$\{k\} = \frac{(\text{mol dm}^{-3}\,\text{s}^{-1})}{(\text{mol dm}^{-3})} = \text{s}^{-1}. \tag{2.10}$$

As a final example, consider the rate law

$$\text{rate} = k[\text{CH}_3\text{CHO}]^{3/2}. \tag{2.11}$$

Substituting in the units for the rate and concentrations gives

$$(\text{mol dm}^{-3}\,\text{s}^{-1}) = \{k\}(\text{mol dm}^{-3})^{3/2} \tag{2.12}$$

which we can solve to give

$$\{k\} = \frac{(\text{mol dm}^{-3}\,\text{s}^{-1})}{(\text{mol dm}^{-3})^{3/2}} = \text{mol}^{-1/2}\,\text{dm}^{3/2}\,\text{s}^{-1}. \tag{2.13}$$

An important point to note is that while we can compare reaction *rates* for reactions with different orders, it makes little sense to try and compare the corresponding rate constants, which have different dimensions (as an analogy, you would not try to compare 4 m with 10 kg).

2.4 Integrated rate laws

A rate law is a differential equation that describes the rate of change of a reactant or product concentration with time. If we integrate the rate law, then we obtain an expression for the concentration as a function of time. This is generally the type of data obtained in an experiment, allowing a direct comparison between the rate law and experimental data. In many simple cases, the rate law may be integrated analytically. Otherwise, numerical (computer-based) techniques may be used. Four of the simplest rate laws are given in table 2.1 in both their differential and integrated form.

2.5 Half lives

The half life, $t_{1/2}$, of a substance is defined as the time it takes for the concentration of the substance to fall to half of its initial value. We can also define consecutive half

Table 2.1. Integrated rate laws for zeroth, first, and second-order reactions. $[A]_0$ and $[B]_0$ represent the initial concentrations of A and B.

Reaction	Order	Differential form	Integrated form
$A \longrightarrow P$	zeroth	$\frac{d[A]}{dt} = -k$	$[A] = [A]_0 - kt$
$A \longrightarrow P$	first	$\frac{d[A]}{dt} = -k[A]$	$\ln[A] = \ln[A]_0 - kt$
$A + A \longrightarrow P$	second	$\frac{1}{2}\frac{d[A]}{dt} = -k[A]^2$	$\frac{1}{[A]} = \frac{1}{[A]_0} + 2kt$
$A + B \longrightarrow P$	second	$\frac{d[A]}{dt} = -k[A][B]$	$kt = \frac{1}{[B]_0 - [A]_0}\ln\frac{[B]_0[A]}{[A]_0[B]}$

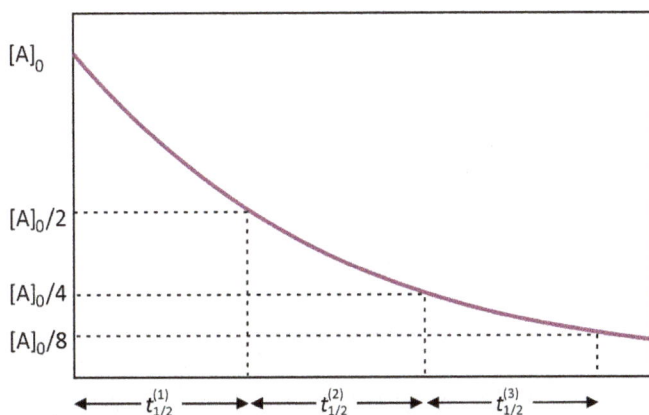

Figure 2.1. The half life of a substance A is the time taken for the concentration of the substance to fall to half of its initial value. Successive half-lives can be defined, as shown.

lives, i.e. the time it takes to fall from half to a quarter of the initial value, from a quarter to an eighth, and so on. This is illustrated in figure 2.1. Note that it only makes sense to define a half life for a substance not present in excess at the start of the reaction. We can obtain equations for the half lives for reactions of various orders by substituting the values $t = t_{1/2}$ and $[A] = [A]_0/2$ into the integrated rate laws given in table 2.1. We obtain

Zeroth order reaction $\quad t_{1/2} = \dfrac{[A]_0}{2k}$

First order reaction $\quad t_{1/2} = \dfrac{ln2}{k}$

Second order reaction $\quad t_{1/2} = \dfrac{1}{k[A]_0}$

By inspection of the expressions for the half lives for reactions of different orders, we see that for a zeroth-order reaction the half life is directly proportional to the initial concentration of the reactant, and that consecutive half lives will therefore decrease by a factor of two. For first-order reactions, the half life has no dependence on the reactant concentrations, and consecutive half lives will be identical. For a second order reaction, the half life is inversely proportional to the initial reactant concentration, with the consequence that consecutive half lives will double. These observations provide a quick way to determine the reaction order from a plot of reaction concentration versus time.

Having introduced the reader to rate laws and a number of related concepts, in the next chapter we will look at methods for determining reaction orders, rate constants, and rate laws from experimental data of the type acquired in a kinetics study.

IOP Concise Physics

An Introduction to Chemical Kinetics

Claire Vallance

Chapter 3

Determining the rate law and obtaining mechanistic information from experimental data

A kinetics experiment consists of measuring the concentrations of one or more reactants or products at a number of different times after initiation of the reaction. This information turns out to be extremely powerful in two contexts:

1. Determining the absolute rate of the reaction and/or the individual elementary steps in the mechanism.

2. Investigating the mechanism of the reaction by comparing the experimental data with rate laws predicted for different reaction mechanisms.

We will review some of the experimental techniques used to make kinetics measurements in chapter 4. The way in which the experimental data will be analysed often has a considerable impact on the experimental design. In the present chapter, we will look at a number of general strategies for analysing data from a kinetics study, and the measurements that must be made in order to employ each strategy. Kinetic data may be analysed to determine the reaction rate, orders with respect to each reactant, and the rate constant for the reaction. As we shall see, these allow the rate law to be elucidated. If measurements are made at a number of temperatures, then the activation energy and pre-exponential factor appearing in the Arrhenius equation may also be determined, often providing further mechanistic information.

3.1 Isolation method

The isolation method is an approach to performing experiments that simplifies the rate law in order to determine its dependence on the concentration of a single reactant at a time. The approach is very simple: the dependence of the reaction rate on the chosen reactant concentration is isolated by having all other reactants present

doi:10.1088/978-1-6817-4664-7ch3

© Morgan & Claypool Publishers 2017

in a large excess, so that their concentration remains essentially constant throughout the course of the reaction. This is best illustrated by means of an example. Consider a reaction, $A + B \longrightarrow P$, in which B is present initially at a concentration 1000 times greater than A. When all of species A has been used up, the concentration of B will only have changed by one part in 1000, or 0.1%, and so 99.9% of the original B will still be present. It is therefore a good approximation to treat its concentration as constant throughout the reaction. The rate law for the reaction then becomes

$$\begin{aligned} \text{rate} &= k[A]^a[B]^b \\ &\approx k[A]^a[B]_0^b \\ &= k_{\text{eff}}[A]^a \end{aligned} \qquad (3.1)$$

with $[B]_0$ the initial concentration of B, and $k_{\text{eff}} = k[B]_0^b$. As we see, the isolation method greatly simplifies the rate law, since the (constant) concentrations of all reactants present in large excess may be combined with the rate constant to yield a single effective rate constant. The new, simplified, rate law only depends on the concentration of one reactant and the order with respect to that reactant. When the rate law depends on the contributions of a number of reactants, a series of experiments may be carried out in which each reactant is isolated in turn.

Rate laws that have been simplified using the isolation method are often referred to in terms of a 'pseudo-order' terminology. For example, if the order with respect to reactant A in the above rate law was 1, the rate law would take the form

$$\text{rate} = k_{\text{eff}}[A],$$

and would be described as *pseudo-first order*. Another way of stating this is to say that the reaction is carried out under 'pseudo-first order conditions'. This means that, under the conditions employed (all reactants, apart from A, are present in large excess), the reaction behaves as if it has first order kinetics, even though the true, unsimplified rate law might have a different order. If the order with respect to [A] in the above example had been 2, the simplified rate law would be said to be *pseudo-second order*, and so on.

Once the rate law has been simplified using the isolation method, the differential or integral methods discussed in the following sections may be used to determine the reaction orders.

3.2 Differential methods: determining reaction orders from the differential form of the rate law

When we have a rate law that depends only on the concentration of one species, either because there is only a single species reacting, or because we have used the isolation method to manipulate the rate law, then the rate law may be written as

$$\text{rate} = k[A]^a. \qquad (3.2)$$

Taking logs of both sides gives

$$\log(\text{rate}) = \log k + a \log[A]. \qquad (3.3)$$

A plot of log(rate) against log[A] will then be a straight line with a slope equal to the reaction order, a, and an intercept equal to $\log k$. There are two ways in which to obtain data to plot in this way:

1. We can measure the concentration of the reactant, [A], as a function of time and use this data to calculate the rate, $-d[A]/dt$, as a function of [A]. A plot of log(rate) against log[A] then yields the reaction order with respect to A. This method has the advantage that all of the data may be acquired in a single experiment; however, it can become difficult to interpret the rate data if secondary reactions occur.

2. We can make a series of measurements of the initial rate of the reaction when different initial reactant concentrations $[A]_0$ are employed. These may then be plotted as above to determine the order, a. This is a commonly used technique known as the *initial rates method*. This approach has the advantage that, because the rate measurement is made at the start of the reaction, secondary reactions will not influence the measurement. However, it does require several experiments to be carried out with different initial concentrations of each reactant, while the other initial reactant concentrations are held constant.

3.3 Integral methods: determining reaction orders from the integrated form of the rate law

Since in an experiment we measure concentrations as a function of time, it is perhaps most natural to compare the data with the appropriate *integrated* rate law (see section 2.4), which also expresses concentration as a function of time. Again, this is most straightforward if we have simplified the rate law, so that it depends on only one reactant concentration. Integrated rate laws for reactions of various orders were given in table 2.1. These can easily be compared with experimental data by employing the appropriate straight-line plots. The most commonly encountered cases are:

1. Zeroth order integrated rate law, $[A] = [A]_0 - kt$
   A plot of [A] versus t will be linear, with a slope of $-k$.

2. First order integrated rate law, $\ln[A] = \ln[A]_0 - kt$
   A plot of $\ln[A]$ versus t will be linear, with a slope of $-k$.

3. Second order integrated rate law, $\frac{1}{[A]} = \frac{1}{[A]_0} + 2kt$
   A plot of $\frac{1}{[A]}$ versus t will be linear, with a slope of $2k$.

If none of these plots result in a straight line, then the reaction is not zeroth, first, or second order, and more complicated integrated rate laws must be tried.

The integral method provides a straightforward way to determine the reaction order. Since the rate constant is determined from the slope of a linear plot, the method often also allows a more accurate determination of the rate constant than the differential method described in the previous section, which relies on a determination of $\log k$ from the intercept.

3.4 Determining orders and rate constants from a half life analysis

Another way of determining the reaction order is to investigate the behaviour of the half life as the reaction proceeds. We have already observed in section 2.5 that the half lives for zeroth, first, and second order reactions have markedly different dependences on the initial reactant concentration.

To perform a half-life analysis, we determine a series of successive half lives from a plot of reactant concentration versus time. The initiation of the reaction, $t = 0$, is used as the start time from which to measure the first half life, $t_{1/2}^{(1)}$. Then, $t = t_{1/2}^{(1)}$ is used as the start time from which to measure the second half life, $t_{1/2}^{(2)}$, and so on. The behaviour of the half lives for zeroth, first, and second order reactions are summarised below:

1. Zeroth order: $t_{1/2} = \frac{[A]_0}{2k}$
   Since at $t = t_{1/2}^{(1)}$, the new starting concentration is $[A]_0/2$, successive half lives will decrease by a factor of two for a zeroth order reaction.

2. First order: $t_{1/2} = \ln 2k$
   There is no dependence of the half life on concentration, so the half life is constant for a first order reaction.

3. Second order: $t_{1/2} = \frac{1}{k[A]_0}$
   The inverse dependence on concentration means that successive half lives will double for a second order reaction.

3.5 Some examples

3.5.1 A decarboxylation reaction: RCOOH $\longrightarrow$ R–H + CO$_2$

In order to illustrate some of the methods outlined above, we will perform a number of different analyses on a set of data from a kinetics study on a generic decarboxylation reaction. The measurement involved tracking the volume of CO_2 evolved from 0.01 dm^3 of a 0.01 mol dm^{-3} solution of the acid, RCOOH, as a function of time. The measured data are shown in table 3.1, and were obtained at atmospheric pressure and a temperature of 298 K.

Table 3.1. Experimental data set from a kinetics measurement on the RCOOH decarboxylation reaction.

t/s	Volume of CO_2 evolved/cm^3
0	0.00
1500	0.64
3000	1.10
4500	1.45
6000	1.70
7500	1.90
9000	2.09
∞	2.45

Table 3.2. Conversion of CO_2 product volume into RCOOH reactant concentration as a function of time for the RCOOH decarboxylation reaction.

t/s	$V(CO_2)$/cm^3	$n(CO_2)$/mol	$n(RCOOH)$/mol	[RCOOH]/mol dm^{-3}
0	0	0	1.00×10^{-4}	1.00×10^{-2}
1500	0.64	2.62×10^{-5}	7.38×10^{-5}	7.38×10^{-3}
3000	1.10	4.50×10^{-5}	5.50×10^{-5}	5.50×10^{-3}
4500	1.45	5.93×10^{-5}	4.07×10^{-5}	4.07×10^{-3}
6000	1.70	6.95×10^{-5}	3.05×10^{-5}	3.05×10^{-3}
7500	1.90	7.77×10^{-5}	2.23×10^{-5}	2.23×10^{-3}
9000	2.09	8.38×10^{-5}	1.62×10^{-5}	1.62×10^{-3}
∞	2.45	1.00×10^{-4}	0	0

If the experiment had tracked a reactant concentration as a function of time, then we could immediately begin to analyse the kinetics. However, in the present example, it is a product concentration that has been tracked, via a measurement of the volume of CO_2 generated in the reaction. All of the methods described above require knowledge of the reactant concentration, [RCOOH], as a function of time, so at first sight we appear to have a problem. In fact, this is a relatively simple problem to solve, since it is very straightforward to convert between product volume and reactant concentration. We begin by determining the number of moles n of the reactant, RCOOH, present initially. This is simply the product of the concentration c and volume V of our starting solution:

$$n = cV = (0.01 \ \text{mol dm}^{-3}) \, (0.01 \ \text{dm}^3) = 1.0 \times 10^{-4} \ \text{mol}. \tag{3.4}$$

Treating the CO_2 product as an ideal gas, we can use the ideal gas law to convert from volume, V, to the number of moles of product, n, given the pressure, p, and temperature, T.

$$n = \frac{pV}{RT}. \tag{3.5}$$

Since one molecule of RCOOH reactant yields one molecule of CO_2 product, the amount of reactant remaining at each time is simply $n(\text{RCOOH}) = n(\text{RCOOH})_0 - n(CO_2)$. To convert this to a concentration, we divide through by the volume of the reaction mixture, 0.01 dm^3. The various steps in this calculation are shown in table 3.2.

Now that we have determined the reactant concentration as a function of time, we can perform a kinetic analysis on the data. We will explore three different ways of determining the reaction order and rate constant, using differential and integral rate laws and a half life analysis.

Integrated rate law method
We can use the integrated rate law method to test a number of rate laws. For example, if the reaction is first order, with a rate equation

$$\frac{d[RCOOH]}{dt} = -k[RCOOH] \qquad (3.6)$$

then the corresponding integrated rate law is

$$[RCOOH] = [RCOOH]_0 e^{-kt} \qquad (3.7)$$

or in log form

$$\ln[RCOOH] = \ln[RCOOH]_0 - kt. \qquad (3.8)$$

A plot of $\ln[RCOOH]$ against t should therefore be linear, with a slope of $-k$.

If the reaction is second order, with a rate law

$$\frac{d[RCOOH]}{dt} = -k[RCOOH]^2 \qquad (3.9)$$

then the corresponding integrated rate law is

$$\frac{1}{[RCOOH]} = \frac{1}{[RCOOH]} + kt. \qquad (3.10)$$

In this case, a plot of $1/[RCOOH]$ against t is expected to be linear, with a slope equal to the rate constant, k.

These two plots are shown in figures 3.1(a) and (b), with the relevant tabulated data given in table 3.3. We see immediately that the plot of the first-order integrated rate law is linear, while that for the second-order rate law is not. The reaction is therefore first order. The slope of the line yields a first-order rate constant $k = 2.01 \times 10^{-4} \text{ s}^{-1}$.

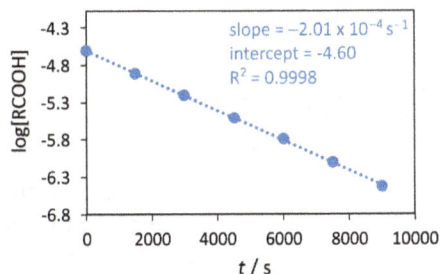

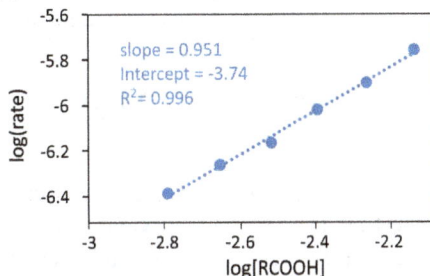

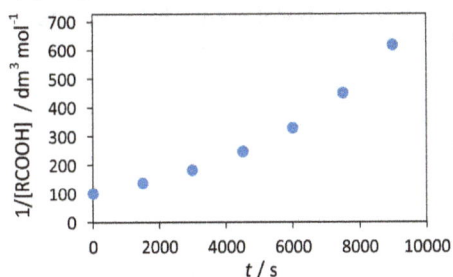

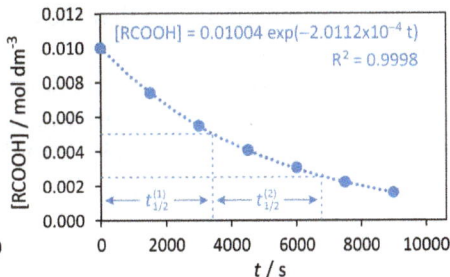

Figure 3.1. Kinetic analysis of a decarboxylation reaction, $RCOOH \longrightarrow R\text{–}H + CO_2$. See text for details.

Table 3.3. Data for first and second order integrated rate law plots and differential rate law plots for the RCOOH decarboxylation reaction, derived from the data in table 3.1.

t/s	ln [RCOOH]	1/[RCOOH]
0	−4.6052	100.00
1500	−4.9085	135.44
3000	−5.2026	181.74
4500	−5.5038	245.61
6000	−5.7929	327.95
7500	−6.1051	448.13
9000	−6.4264	617.97

Differential method

Alternatively, we can use the differential form of the rate law to perform our analysis. Starting from a rate law with an unknown order,

$$\text{rate} = -\frac{d[\text{RCOOH}]}{dt} = k[\text{RCOOH}]^a \tag{3.11}$$

we can take logs of both sides to obtain

$$\log(\text{rate}) = \log k + a \log [\text{RCOOH}]. \tag{3.12}$$

A plot of log(rate) against log[RCOOH] should therefore be linear with a slope equal to the reaction order, a, and an intercept equal to $\log k$. The rate associated with each data point in table 3.2 is determined numerically from the data in the table using

$$\text{rate} = \frac{d[\text{RCOOH}]}{dt} \approx \frac{\Delta[\text{RCOOH}]}{\Delta t} = \frac{[\text{RCOOH}]_i - [\text{RCOOH}]_{i-1}}{t_i - t_{i-1}} \tag{3.13}$$

where $[\text{RCOOH}]_i$ and t_i denote the reactant concentration and time at data point i. We note that this expression yields the *average* rate within each time period rather than the instantaneous rate at each measurement time, which does introduce some error into the analysis. The results of the rate determination are shown in table 3.4.

The plot of equation (3.12) is shown in figure 3.1(c), and is indeed linear. The slope is close to unity, implying that the reaction is first order, as we found using the integrated rate law method above. The intercept yields $\log k = -3.74$, or $k = 1.80 \times 10^{-4}$. This is close to the value for the rate constant determined above from the integrated rate law plot, but is somewhat less reliable, due to errors introduced when using equation (3.13) to determine the rate from the experimental data.

Half life analysis

Finally, we can determine consecutive half lives for the RCOOH reactant from a plot of [RCOOH] against time, as shown in figure 3.1(d). We see that the first and

Table 3.4. Data for a differential rate law plot for the RCOOH decarboxylation reaction.

t/s	$[RCOOH]/mol\ dm^{-3}$	rate/mol $dm^{-3}\ s^{-1}$	log [RCOOH]	log (rate)
0	1.000×10^{-2}	–	−2.000	–
1500	7.383×10^{-3}	1.744×10^{-6}	−2.132	−5.758
3000	5.503×10^{-3}	1.254×10^{-6}	−2.259	−5.902
4500	4.071×10^{-3}	9.540×10^{-7}	−2.390	−6.020
6000	3.049×10^{-3}	6.815×10^{-7}	−2.516	−6.167
7500	2.232×10^{-3}	5.452×10^{-7}	−2.651	−6.263
9000	1.618×10^{-3}	4.089×10^{-7}	−2.791	−6.388

second half lives are equal, implying that the reaction is first order, in agreement with our results from the integral and differential methods employed previously. For a first-order reaction, we can fit the exponentially decaying concentration to the rate law given in equation (3.7). In the present case, the fit was performed within Excel using the 'trendline' feature. We obtain an excellent fit, which predicts the rate constant to be $k = 2.01 \times 10^{-4}\ s^{-1}$, in perfect agreement with the rate law determined above from a plot of the integrated first-order rate law.

3.5.2 Determining the rate law for the reaction CO + Cl₂ ⟶ COCl + Cl using the method of initial rates

As a second short example, we will consider a data set collected for the reaction between CO and Cl_2 using the initial rates method. Table 3.5 shows the initial rate of the reaction, as determined from the slope of a concentration versus time plot, for a number of different initial concentrations of the two reactants.

The rate law for the reaction, expressed in terms of reactant pressures rather than concentrations, takes the form

$$\text{rate} = k p_{Cl_2}^a p_{CO}^b \tag{3.14}$$

where a and b are the orders to be determined. When the initial pressure of each reactant is held constant in turn and the initial rate is measured, we can take logs of both sides to obtain:

$$\log (\text{initial rate}) = \log k p_{Cl_2,0}^a + b \log p_{CO}$$
$$\log (\text{initial rate}) = \log k p_{CO,0}^b + a \log p_{Cl_2} \tag{3.15}$$

where $p_{x,0}$ denotes the initial pressure of the reactant , x, whose initial pressure is held constant. We therefore expect that in the case where the initial concentration of Cl_2 is held constant, a plot of log(initial rate) against $\log p_{CO}$ should yield a straight line with slope b, the order with respect to p_{CO}. Similarly, when the initial pressure of CO is held constant, a plot of log(initial rate) against $\log p_{Cl_2}$ should yield a straight line with slope a, the order with respect to p_{Cl_2}. These plots are shown in figure 3.2.

Table 3.5. Rate data for the reaction $CO + Cl_2 \longrightarrow COCl + Cl$ from a study employing the method of initial rates.

Initial p_{Cl_2}/Torr	Initial p_{CO}/Torr	Initial rate/Torr s^{-1}	$\log(p_{Cl_2})$	$\log(p_{CO})$	$\log(\text{rate})$
665	88	1.44	2.82	1.94	0.158
480	88	0.88	2.68	1.94	−0.056
230	88	0.29	2.36	1.94	−0.538
90	640	0.52	1.95	2.81	−0.284
90	465	0.38	1.95	2.67	−0.420
90	210	0.17	1.95	2.32	−0.770

Straight lines are obtained in both cases, and yield an order of 3/2 with respect to p_{Cl_2} and 1 with respect to p_{CO}. The rate law is therefore

$$\text{rate} = kp_{Cl_2}^{3/2}p_{CO}. \tag{3.16}$$

Despite the simple form of the rate law, the fractional order implies that the reaction proceeds via a complex (multi-step) mechanism.

In addition to determining the rate law, we can use the intercept of either plot to find the rate constant, k. From equation (3.15) and the intercepts of the two plots, we have

$$\log(kp_{CO,0}) = -4.1021$$
$$\log(kp_{Cl_2,0}) = -3.1027. \tag{3.17}$$

Substituting the relevant initial pressures into these equations yields a rate constant of $k = 8.98 \times 10^{-7}$ Torr $^{-3/2}$ s^{-1} from the first plot and $k = 9.25 \times 10^{-7}$ Torr $^{-3/2}$ s^{-1} from the second plot. The two values are in agreement to within 3%.

3.6 Exploring the temperature dependence of the rate constant

As noted previously in section 1.4.6, it is found experimentally that the dependence of the rate constant on temperature, T, for many chemical reactions follows the Arrhenius equation,

$$k(T) = A\exp\left(-\frac{E_a}{RT}\right) \tag{3.18}$$

or in log form

$$\ln k = \ln A - \frac{E_a}{RT} \tag{3.19}$$

where A is the pre-exponential factor and E_a is the activation energy. The pre-exponential factor and activation energy are known collectively as the *Arrhenius parameters*. If the rate constant for a reaction is measured at a number of different temperatures, the Arrhenius parameters may be determined from a plot of $\ln k$ against $1/T$. The resulting *Arrhenius plot* has an intercept of $\ln A$ and a slope

(a)

(b)

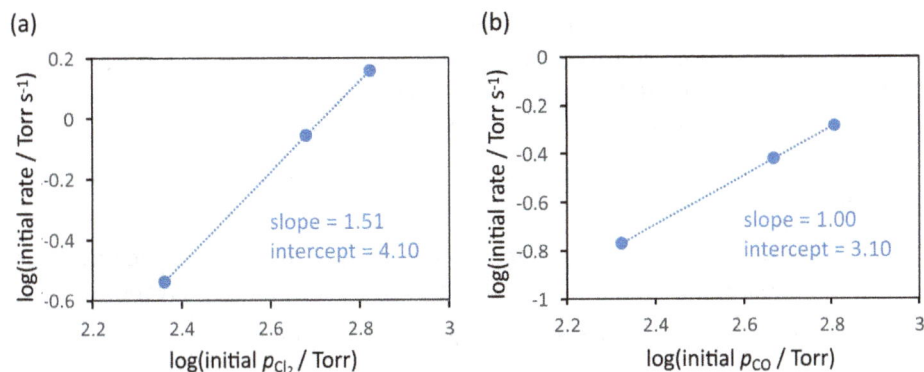

Figure 3.2. Differential rate law plots for the reaction $CO + Cl_2 \longrightarrow COCl + Cl$.

of $-E_a/R$. For most reactions, the Arrhenius equation works fairly well over at least a limited temperature range. However, there are often deviations. These may be due to the fact that the pre-exponential is not in fact a constant, but has its own temperature dependence[1], or due to more exotic effects, such as a contribution to the rate from quantum tunnelling. Tunnelling through the activation barrier commonly makes a significant contribution to the reaction rate for proton or electron transfer reactions at low temperatures[2]. When the reaction tunnels through the barrier, rather than surmounting it, the 'effective' activation barrier is drastically lowered, with a concominant increase in the reaction rate.

For an elementary reaction, both the activation energy and the pre-exponential factor have definite physical interpretations. In particular, the activation energy may be interpreted as the energy difference between the reactants and the transition state involved in the chemical rearrangement (see figure 1.2). When the Arrhenius equation is applied to the overall kinetics of a multi-step reaction, E_a simply becomes an experimental parameter describing the temperature dependence of the overall reaction rate. E_a may vary with temperature, and may take positive or negative values. In this context, we can define the activation energy as:

$$E_a = RT^2 \frac{d \ln k}{dT} = -R \frac{d \ln k}{d(1/T)}.$$ (3.20)

This is a more general definition of the activation energy than the Arrhenius equation, with the two definitions becoming equivalent in the case when E_a is independent of temperature (all you need to do to prove this is to integrate the above

---

[1] An investigation into the detailed temperature dependence of A is beyond the scope of this book. The behaviour is a consequence of the thermal distribution of quantum states in the reacting molecules at different temperatures, and can be treated rigorously within the framework of statistical mechanics and transition-state theory.

[2] The probability of tunnelling depends on the height and width of the barrier, and on the mass of the tunnelling particle. The tunnnelling rate is highest for reactions with low, narrow barriers, and involving the transfer of very light particles. For this reason, it is most commonly observed in reactions involving electron or proton transfer.

equation, treating E_a as a constant). With the above definition, we can determine E_a at a given temperature from the slope (at the temperature of interest) of a plot of $\ln k$ against $1/T$, even if the Arrhenius plot is not a straight line.

There are a few observations that follow from equation (3.20)

1. The higher the activation energy, the stronger the temperature dependence of the rate constant.

2. A reaction with no temperature dependence has an activation energy of zero (this is common in ion–molecule reactions and radical–radical recombinations).

3. A negative activation energy implies that the rate decreases as the temperature increases, and always indicates a complex reaction mechanism. An example of a reaction with a negative activation energy is the oxidation of NO to form NO_2, which has the mechanism

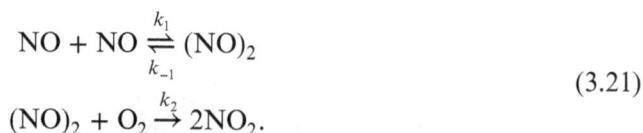

$$NO + NO \underset{k_{-1}}{\overset{k_1}{\rightleftharpoons}} (NO)_2$$

$$(NO)_2 + O_2 \overset{k_2}{\rightarrow} 2NO_2. \tag{3.21}$$

At higher temperatures, the intermediate complex, $(NO)_2$, becomes more unstable and has a shorter lifetime. There is less time for the O_2 molecule to react with the complex in order to form the NO_2 products, and the reaction rate therefore decreases. Another way of thinking about this is that formation of the complex is exothermic, and increasing the temperature will therefore shift the pre-equilibrium to the left (by Le Chatelier's principle), again reducing the overall rate of reaction.

3.6.1 Overall activation energies for complex reactions

When dealing with complex reactions, the Arrhenius equation can often be used to estimate the overall activation energy from knowledge of the activation energies of the individual steps. For example, in the above reaction, the overall rate law (you will learn how to derive a rate law of this type in section 6.4) is

$$\frac{1}{2}\frac{d[NO_2]}{dt} = \frac{k_1 k_2}{k_{-1}}[NO]^2[O_2] = k[NO]^2[O_2] \tag{3.22}$$

where $k = k_1 k_2 / k_{-1}$ is the observed third-order rate constant. The temperature dependence of k can be related to the activation barriers for each elementary step by applying the Arrhenius equation to each of k_1, k_{-1}, and k_2, as follows.

$$k = \frac{k_1 k_2}{k_{-1}}$$

$$= \frac{A_1 \exp\left(-\frac{E_{a,1}}{RT}\right) A_2 \exp\left(-\frac{E_{a,2}}{RT}\right)}{A_{-1} \exp\left(-\frac{E_{a,-1}}{RT}\right)} \tag{3.23}$$

$$= \frac{A_1 A_2}{A_{-1}} \exp\left(\frac{-E_{a,1} - E_{a,2} + E_{a,-1}}{RT}\right).$$

From this expression, we can identify the Arrhenius parameters for the overall reaction:

$$A = \frac{A_1 A_2}{A_{-1}}$$

$$E_a = - E_{a,1} - E_{a,2} + E_{a,-1}.$$

(3.24)

3.7 Summary

In the present chapter, we have considered in some detail how various kinetic parameters of interest may be extracted from data sets consisting of reactant or product concentrations as a function of time after initiation of the reaction. In the next chapter, we will investigate the general requirements that must be kept in mind when recording such data sets, and will explore a variety of methods suitable for acquiring kinetic data across a broad range of timescales.

Chapter 4

Experimental techniques for measuring reaction rates

A wide variety of experimental techniques have been developed in order to monitor chemical reaction rates over timescales ranging from hours or days all the way down to a few femtoseconds. While it is relatively simple to monitor the kinetics of a slow reaction (proceeding over a timescale of minutes to hours or longer), highly specialised techniques are required in order to study fast reactions. We will look at a variety of methods that collectively span the entire range of timescales of interest.

Whatever the details of the experimental arrangement, any kinetics experiment essentially consists of mixing the reactants and initiating the reaction on a timescale that is negligible relative to that of the reaction, and then monitoring the concentration(s) of one or more reactants and/or products as a function of time. Because rate constants vary with temperature, it is also important to monitor and control the temperature at which the reaction occurs.

Most of the techniques we will look at are batch techniques, in which the reaction is initiated at a single chosen point in time, and concentrations are then followed as a function of time after initiation. We will also consider one or two examples of continuous techniques, in which the reaction is continuously initiated and the time dependence of the reaction mixture composition is inferred from, for example, the concentrations in different regions of the reaction vessel. The continuous flow method outlined in the next section is an example of such a technique.

4.1 Techniques for mixing the reactants and initiating reaction

For slow reactions, occurring over minutes to hours, the reaction is usually initiated simply by mixing the reactants together by hand or with a magnetic stirrer or other mechanical device. For fast reactions, a wide range of techniques have been developed.

doi:10.1088/978-1-6817-4664-7ch4

© Morgan & Claypool Publishers 2017

4.1.1 Flow techniques

Flow techniques are typically used to study reactions occurring on timescales of seconds to milliseconds. In the simplest flow method, shown schematically in figure 4.1(a), reactants are mixed at one end of a flow tube, and the composition of the reaction mixture is monitored at one or more positions further along the tube. If the (constant) flow velocity v_{flow} along the tube is known, then measurements at different positions x provide information on concentrations at different times, $t = x/v_{flow}$, after initiation of the reaction. In a variation on this method, shown in figure 4.1(b), the position of the detector is fixed, but a moveable injector is used to inject one of the reactants into the flow tube. By injecting the reactant at different positions relative to the detector, the time dependence of the reaction mixture composition can be studied. Reactions of atomic or radical species may be studied using the discharge flow method, in which the reactive species is generated by a microwave discharge immediately prior to injection into the flow tube.

Continuous flow methods have the disadvantages that relatively large quantities of reactants are needed, and also that very high flow velocities are required in order to study fast reactions. These problems are addressed in the *stopped flow* technique. In this method, the reactants are rapidly flowed into a fixed-volume reaction chamber, and are mixed by the action of a syringe fitted with an end stop (see figure 4.2). The composition of the reaction mixture is then monitored spectroscopically as a function of time after mixing at a fixed position in the reaction chamber. Experimental systems have been designed to allow measurements to be made on

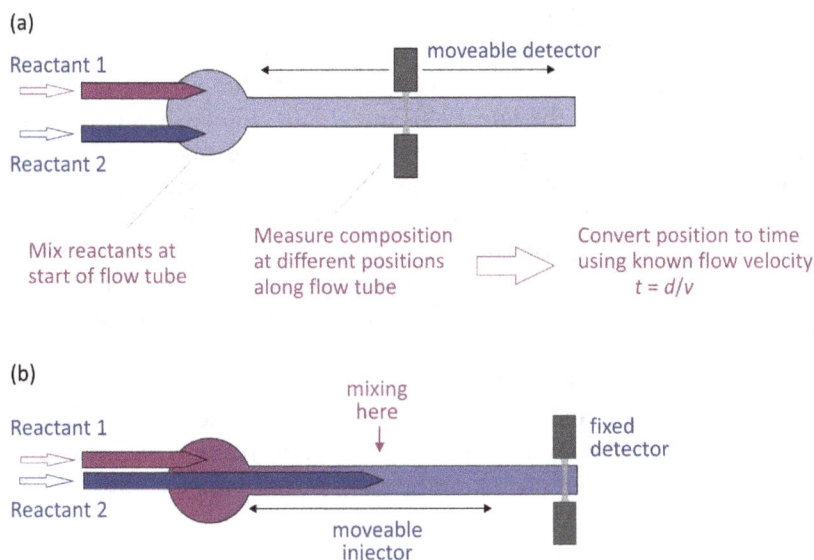

Figure 4.1. Schematic of the basic principles of flow techniques as employed in kinetics measurements: (a) reactants are mixed, the reaction mixture is flowed along a flow tube at a constant flow velocity, and the mixture is sampled at different positions along the flow tube by a moveable detector, in order to determine the composition of the mixture as a function of time after initiation of the reaction; (b) instead of moving the detector, the position at which one of the reactants is injected into the flow tube can be adjusted instead.

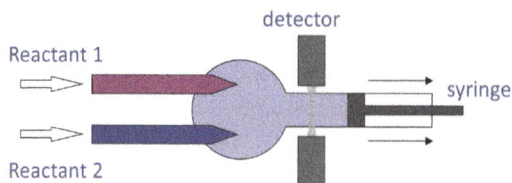

Figure 4.2. Schematic of a stopped flow experiment. Reactants are mixed, and a fixed volume of the reaction mixture is drawn through the flow tube and analysed as a function of time.

very small sample volumes, making the stopped flow method popular for studying the kinetics of biochemical reactions, for example enzyme-catalysed reactions of the type described in section 6.5. All flow techniques share the common problem that contributions from heterogeneous reactions at the walls of the flow tube can complicate the experiments. Surface effects can be minimised by coating the inner surface of the flow tube with an unreactive substance, such as teflon or halocarbon wax. Alternatively, the relative contributions from the process under study and reactions involving the walls may be quantified by varying the diameter of the flow tube, and therefore the ratio of volume to surface area.

4.1.2 Flash photolysis and laser pump probe techniques

In the flash photolysis method, the reaction is initiated by a pulse of light (the 'flash'), which dissociates a suitable precursor molecule in the reaction mixture to produce a reactive species. The concentration of the reactive species is then monitored as a function of time, usually via a spectroscopic technique (these will be discussed in section 4.2.1). The shortest timescale over which reactions may be studied is determined by the duration of the flash. In the earliest flash photolysis experiments, carried out in the late 1940s and 1950s, the flash was provided by a discharge lamp, with durations in the region of tens of microseconds to several milliseconds. In most modern experiments, the flash is provided by a laser pulse, typically with a duration of a few nanoseconds ($1\ ns=10^{-9}$ s). For studying extremely fast reactions, such as some of the electron transfer processes involved in photo-synthesis, laser pulses as short as a few tens of femtoseconds ($1\ fs=10^{-15}$ s) may be employed. Pulse radiolysis is a variation on flash photolysis in which a short pulse of high energy electrons, from nanoseconds to microseconds in duration, is passed through the sample in order to initiate a reaction.

Both flash photolysis and pulse radiolysis have the advantage that, because reactants are produced from well-mixed precursors, there is no mixing time to reduce the time resolution of the technique. Also, because the reactants are generated and monitored in the centre of the reaction cell, there are no wall reactions to complicate the data analysis.

For very fast processes, the *laser pump–probe* technique is often used, in which pulsed lasers are employed both to initiate the reaction (the 'pump') and to detect the products via a pulsed spectroscopic technique (the 'probe'). The time separation between the two pulses can be varied either electronically or with an optical delay line, down to a resolution of around 10 fs (10^{-14} s)

4.1.3 Relaxation methods

If we allow a chemical system to come to equilibrium and then perturb the equilibrium in some way, the rate of relaxation to a new equilibrium position provides information about the forward and reverse rate constants for the reaction. Since a system at chemical equilibrium is already well-mixed, relaxation methods overcome the mixing problems associated with many flow methods.

As an example, we will investigate the effect of a sudden increase in temperature on a system at equilibrium, an experiment known as a *temperature jump*. Consider a simple equilibrium

$$A \underset{k_{1b}}{\overset{k_{1f}}{\rightleftharpoons}} B \tag{4.1}$$

where k_{1f} and k_{1r} are the rate constants for the forward and reverse reactions at the initial temperature, T_1. The rate of change of [A] is

$$\frac{d[A]}{dt} = -k_{1f}[A] + k_{1r}[B]. \tag{4.2}$$

At equilibrium, the rates of the forward and reverse reactions are equal, and so

$$k_{1f}[A]_{eq,1} = k_{1b}[B]_{eq,1}. \tag{4.3}$$

We now increase the temperature suddenly by a few degrees. This is often achieved by discharging a high voltage capacitor through the solution, or by employing a UV or IR laser pulse or microwave discharge. After the temperature jump, the concentrations of A and B are initially at the values $[A]_{eq,1}$ and $[B]_{eq,1}$, but the system is not at the equilibrium composition for the higher temperature. The system relaxes to the new equilibrium concentrations $[A]_{eq,2}$ and $[B]_{eq,2}$ at a rate determined by the new higher-temperature rate constants k_{2f} and k_{2r}. This is shown schematically in figure 4.3. The new concentrations are given by

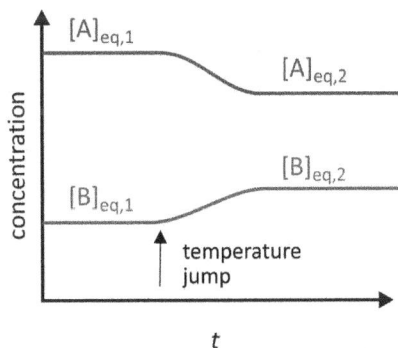

Figure 4.3. When a system at equilibrium, A ⇌ B, undergoes a temperature jump, the concentrations of A and B relax back to new equilibrium values at a rate determined by the rate constants of the forward and backward reactions.

$$k_{2f}[A]_{eq,2} = k_{2b}[B]_{eq,2}. \tag{4.4}$$

We define $x = [A] - [A]_{eq,2}$ as the deviation of the concentration [A] from its new (higher temperature) equilibrium value. The deviation of [B] from its equilibrium value must therefore be $-x$. During the relaxation, the concentrations change as follows

$$
\begin{aligned}
\frac{d[A]}{dt} &= -k_{2f}[A] + k_{2r}[B] \\
&= -k_{2f}([A]_{eq,2} + x) + k_{2r}([B]_{eq,2} - x) \\
&= -(k_{2f} + k_{2r})x
\end{aligned} \tag{4.5}
$$

where we have used equation (4.4) to reach the final result. Since the rate of change of [A] is the same as the rate of change of x, we can integrate the rate law to give

$$x = x_0 \exp(-t/\tau) \tag{4.6}$$

with

$$\frac{1}{\tau} = k_{2f} + k_{2r}. \tag{4.7}$$

We see that the rate at which the concentrations relax to their new equilibrium values is determined by the sum of the two new rate constants. The new equilibrium constant is given by the ratio of the two rate constants, $K_2 = k_{2f}/k_{2r}$, so a measurement of both the rate of relaxation to the new equilibrium and the equilibrium constant allows the individual reaction rate constants for the forward and reverse reaction to be determined.

When employing relaxation methods, the details of the kinetic equations change for more complicated reactions, but the basic principle of the technique remains the same.

4.1.4 Shock tubes

The shock tube method provides a way of producing highly reactive atomic or radical species through rapid dissociation of a molecular precursor, without the use of a discharge or laser pulse. The method is based on the fact that a very rapid increase in pressure (the shock) causes rapid heating of a gas mixture to a temperature of several thousand Kelvin. Since most dissociation reactions are endothermic, at high temperatures their equilibria are shifted towards products. A sudden increase in temperature therefore leads to the rapid production of reactive species (the dissociation products), thereby initiating the reaction of interest. A shock tube, shown schematically in figure 4.4, consists of two chambers separated by a diaphragm. One chamber contains the appropriate mixture of reactants and precursor, the second an inert gas at high pressure. To initiate the reaction, the diaphragm is punctured and a shock wave propagates through the reaction mixture. The temperature rise can be controlled by varying the pressure and composition of

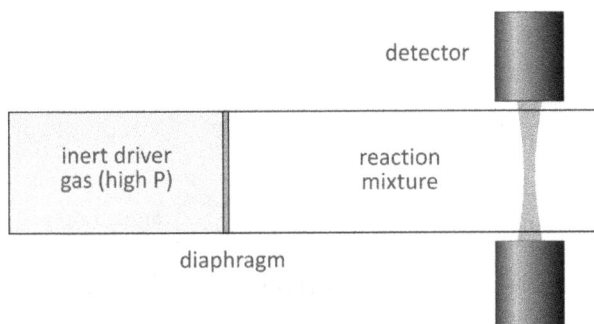

Figure 4.4. A schematic depiction of a shock tube. Rapid heating of the reaction mixture is induced by turbulence generated by a supersonic shock wave after the diaphragm is ruptured.

Table 4.1. Precursors and reactive species employed in shock tube experiments.

$HCN \longrightarrow H + CN$	$CH_4 \longrightarrow CH_3 + H$
$SO_2 \longrightarrow SO + O$	$N_2O \longrightarrow N_2 + O$
$CH_3 \longrightarrow CH_2 + H$	$C_2H_2 \longrightarrow C_2H + H$
$H_2S \longrightarrow HS + H$	$CF_3Cl \longrightarrow CF_3 + Cl$
$NO \longrightarrow N + O$	$C_2H_4 \longrightarrow C_2H_3 + H$
$NH_3 \longrightarrow NH_2 + H$	$C_2H_4 \longrightarrow C_2H_2 + H_2$

the inert gas. The composition of the reaction mixture after initiation is monitored in real time, usually spectroscopically.

The shock tube approach is often used to study combustion reactions. Suitable precursors for such studies, together with the radical species obtained on dissociation using argon as the shock gas, are shown in table 4.1. The shock tube method does have some major drawbacks, not least of which is the fact that the rapid heating is not selective for the formation of particular reactive species, and is likely to lead to at least partial dissociation of all of the molecules in the 'reactants' chamber. This generates a complicated mixture of reactive species, and often a large number of side reactions occur in addition to the reaction of interest. Modelling the kinetics of such a system is often challenging, to say the least. Also, because each experiment is essentially a 'one off', no signal averaging is possible, and signal-to-noise levels are often low. This should be compared with laser pump–probe methods, in which hundreds or even thousands of repeated measurements may be averaged to obtain a good signal-to-noise ratio.

4.1.5 Lifetime methods

A basic principle of quantum mechanics is the Heisenberg uncertainty principle, which relates the uncertainties in the position, x, and momentum, p, of a particle.

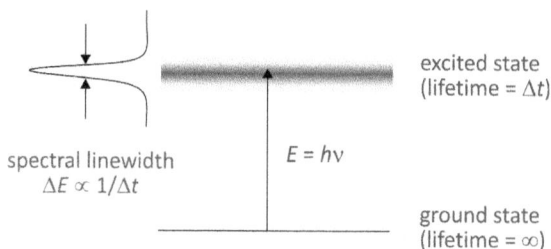

Figure 4.5. The quantum-mechanical uncertainty relationship between energy and time leads to a measureable uncertainty in energy for any atomic or molecular state with a very short lifetime. As a result, the energy of the atomic or molecular ground state tends to be very sharply defined, while the energy of excited states tends to be somewhat 'fuzzy'.

$$\Delta x \Delta p \geqslant h/4\pi \qquad (4.8)$$

where h is Planck's constant. A similar uncertainty principle relates the uncertainties in a particle's energy and lifetime.

$$\Delta E \Delta t \geqslant \frac{h}{4\pi}. \qquad (4.9)$$

The result of this relationship is that an atomic or molecular state has an uncertainty ΔE in its energy that is related to the lifetime Δt of the state. The lifetime of a molecule in its ground state is effectively infinite, so that the uncertainty in the ground-state energy is negligible. However, excited states are short-lived, and there is an associated energy uncertainty, as shown in figure 4.5; the excited state is somewhat 'fuzzy' when drawn on an energy-level diagram. In the context of spectroscopy, photons corresponding to any energy within this uncertainty ΔE may be absorbed, leading to spectral lines having a finite width, known as the natural linewidth.

Kinetic processes involving excited states reduce the lifetime of the state and cause the energy uncertainty to increase, leading to further broadening of spectral lines. Many such processes have first order kinetics, and in these cases the rate constant is simply equal to the reciprocal of the lifetime, $k = 1/\Delta t$. As a consequence, first-order rate constants may be determined directly from measurements of spectral linewidths, provided that other sources of line broadening are absent. Lifetime techniques cover a broad range of timescales, from around 10^{-15} s in photoelectron spectroscopy to around 1 s in nuclear magnetic resonance (NMR) spectroscopy.

4.2 Techniques for monitoring concentrations as a function of time

For slow reactions, the composition of the reaction mixture may be analysed while the reaction is in progress, either by withdrawing a small sample or by monitoring the bulk. This type of approach is generally referred to as a *real-time analysis*. Another option is to use the *quenching method*, in which the reaction is stopped a certain time after initiation so that the composition may be analysed at leisure. Quenching may be achieved in a number of ways. For example:

- sudden cooling;

- adding a large amount of solvent;

- rapid neutralisation of an acid reagent;

- removal of a catalyst;

- addition of a quencher.

The key requirement is that the reaction must be slow enough—or the quenching method fast enough—for little reaction to occur during the quenching process itself. Often, the real-time and quenching techniques are combined by withdrawing and quenching small samples of the reaction mixture at a series of times during the reaction.

The composition of the reaction mixture may be followed in any one of a variety of different ways by tracking any chemical or physical change that occurs as the reaction proceeds. Examples include:

- for reactions in which at least one reactant or product is a gas, the progress of the reaction may be followed by monitoring the pressure in a sealed system, the volume of gas evolved, or the change in mass of the reaction mixture;

- for reactions involving ions, conductivity or pH measurements may often be employed;

- if the reaction is slow enough, the reaction mixture may be titrated;

- if one of the components is coloured, then colourimetry may be appropriate;

- absorption or emission spectroscopy are common. These will be explored in more detail in section 4.2.1;

- for reactions involving chiral compounds, polarimetry (measurement of optical activity) may be useful;

- other techniques include mass spectrometry, gas chromatography, NMR/ESR, and many more.

We now move on to consider fast reactions, which present a much greater experimental challenge. Fast reactions require a fast measurement technique, and as a consequence are usually monitored spectroscopically. Some of the more commonly used spectroscopic techniques are outlined below.

4.2.1 Absorption spectroscopy and the Beer–Lambert law

Absorption spectroscopy is widely used to track reactions in which the reactants and products have different absorption spectra. A simple absorption spectroscopy setup is shown in figure 4.6. A monochromatic light source, often a laser beam, is passed through the reaction mixture, and the ratio of the transmitted to incident light intensity, I/I_0, is measured as a function of time. The quantity $T = I/I_0$ is known as the transmittance,

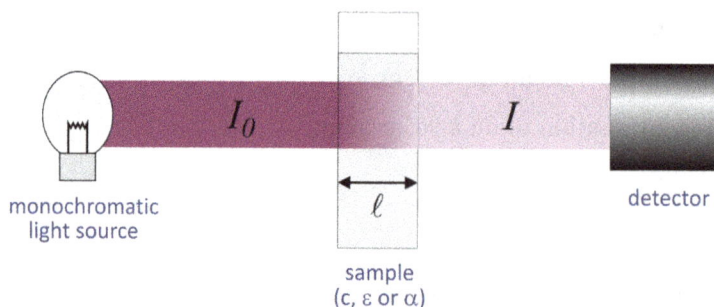

Figure 4.6. Schematic of an absorption spectrometry measurement. Light of intensity I_0 is attenuated to intensity I on passing through a sample of concentration c with absorption/extinction coefficient α/ε present over a pathlength l.

and may be related to the changing concentration of the absorbing species using the Beer–Lambert law, which has both 'decadic' and 'exponential' forms, as follows.

$$T = \frac{I}{I_0} = 10^{-\varepsilon c l} = e^{-\alpha c l} \tag{4.10}$$

where c is the concentration of the absorbing species and l is the path length through the sample. The wavelength-dependent quantities ε and α are known as the molar extinction coefficient and molar absorption coefficient, respectively. These are intrinsic properties of the molecule under study, and are a measure of the efficiency with which the molecule absorbs light at the wavelength or wavelengths of interest. The Beer–Lambert law often appears in logarithmic form:

$$\log(I/I_0) = -\varepsilon c l \tag{4.11}$$

or

$$\ln(I/I_0) = -\alpha c l. \tag{4.12}$$

The quantity $A = \log(I_0/I)$ is known as the *absorbance*, with $A = -\log T$, and is also sometimes referred to as the *optical density*.

Assuming that the relevant absorption or extinction coefficient is known[1], we see from equations (4.10)–(4.12) that if the transmittance or absorbance is measured for a known absorption pathlength, the concentration can be determined.

4.2.2 Resonance fluorescence

Resonance fluorescence is a widely used spectroscopic method for detecting atomic species such as H, N, O, Br, Cl, or F. The experimental setup is shown schematically in figure 4.7.

The light source is a discharge lamp filled with a mixture of helium and a molecular precursor for the atom of interest. A microwave discharge inside the lamp

---

[1] If the extinction coefficient or absorption coefficient is not known, it can often be determined in a separate experiment in which the transmittance or absorbance is measured for various known concentrations of the species of interest.

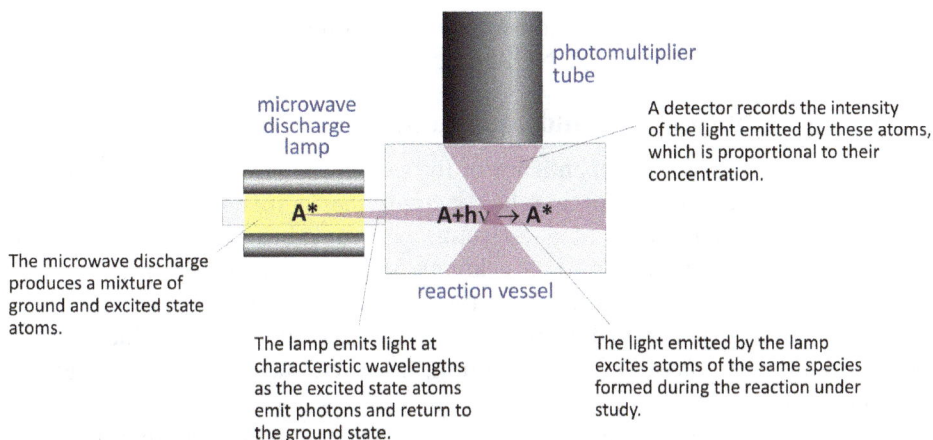

Figure 4.7. Schematic of a resonance fluorescence experiment.

dissociates the precursor and produces a mixture of ground-state and excited-state atoms. As the excited atoms emit photons to return to the ground state, the lamp emits radiation at characteristic frequencies. This radiation may be used to excite atoms of the same species present in a reaction mixture, and monitoring the intensity of radiation emitted from these atoms as they relax back to the ground state provides a measure of their concentration in the reaction mixture. To ensure that the detected light originates from atoms in the reaction mixture and not from atoms within the lamp, the detector—usually a photomultiplier tube—is placed at right angles to the direction in which radiation exits the lamp.

4.2.3 Laser-induced fluorescence

In laser-induced fluorescence (see figure 4.8), a laser is used to excite a chosen species in a reaction mixture to an electronically excited state. The excited states then emit photons to return to the ground state, and the intensity of this fluorescent emission is measured. Because the number of excited states produced by the laser pulse is proportional to the number of ground state molecules present in the reaction mixture, the fluorescence intensity provides a measure of the concentration of the chosen species. This technique is exquisitely sensitive, but does rely on the molecule

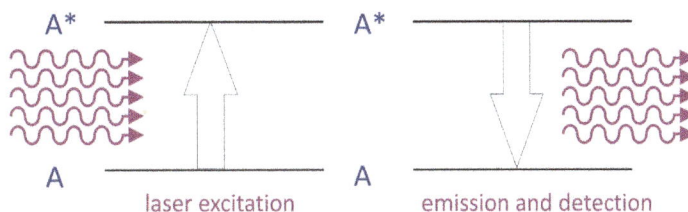

Figure 4.8. The basic principle of laser-induced fluorescence. The amount of emitted light detected after laser excitation is proportional to the concentration of ground-state A in the reaction mixture.

to be detected having a suitable fluorescent excited state. The method is widely used to study reactions involving OH, NO, and other small radical species.

4.3 Temperature control and measurement

The Arrhenius temperature dependence of the rate constant means that, in order to measure an accurate value for a rate constant, the temperature of the reaction mixture must be maintained at a constant, known value. Sometimes activation energies will be measured as part of a kinetic study, in which case rate constants must be measured at a series of temperatures. The temperature of a reaction mixture is most commonly monitored using a thermocouple, due to its wide range of operations and the potential for automation; however, standard thermometers are also commonly used.

There are numerous ways in which the temperature of a reaction mixture may be controlled. For example, reactions in the liquid phase may be carried out in a temperature-controlled thermostat, while reactions in the gas phase are usually carried out inside a stainless steel vacuum chamber, in which thermal equilibrium at the temperature of the chamber is maintained through collisions of the gas molecules with the chamber walls. High temperatures up to 1300 K may be obtained using conventional heaters. Low temperatures may be achieved by flowing cooled liquid through the walls of the reaction vessel, and very low temperatures may be reached by using cryogenic liquids, such as liquid nitrogen (~77 K) or liquid helium (~4 K). For very specialised studies, even lower temperatures of a few Kelvin may be obtained by preparing the reactant gases in a supersonic expansion. Such temperatures are relevant for reactions occurring in environments such as interstellar gas clouds.

IOP Concise Physics

An Introduction to Chemical Kinetics

Claire Vallance

Chapter 5

Introduction to complex reactions

In kinetics, the term 'complex reaction' simply refers to a reaction whose mechanism comprises more than one elementary step. Over the remaining chapters, we will look at a range of different types of complex reactions and the rate laws that may be predicted from their kinetic mechanisms.

In the previous chapters, we have investigated a variety of experimental methods for determining reaction orders and rate constants, and hence establishing the rate law for a given reaction. Such an experimentally determined rate law may be compared with the predicted rate law for a given mechanism in order to determine whether the proposed mechanism is indeed the correct one. Disagreement of a predicted rate law with the experimental data is enough to rule out the corresponding proposed mechanism, while agreement inspires some confidence that the proposed mechanism is correct. It should be noted though that agreement between the predicted and measured kinetics is not always enough to confirm a mechanism. Any proposed mechanism must also be able to account for all other properties of the reaction, which may include quantities such as the product distribution, product stereochemistry, kinetic isotope effects, temperature dependence, and so on.

Over the next few chapters, we will investigate the kinetics of a number of reactions with complex mechanisms. We will begin in the present chapter with two simple examples, namely a mechanism consisting of a number of consecutive or sequential, irreversible elementary steps, and a situation known as a pre-equilibrium. In chapter 6, we will cover unimolecular reactions, third-order reactions, and enzyme reactions, and in chapter 7 we will look into chain reactions and explosions, categories of reaction for which the reaction mechanism can be extremely complex.

5.1 Consecutive reactions

The simplest complex reaction consists of two consecutive, irreversible elementary steps.

doi:10.1088/978-1-6817-4664-7ch5   © Morgan & Claypool Publishers 2017

$$A \xrightarrow{k_1} B \xrightarrow{k_2} C. \tag{5.1}$$

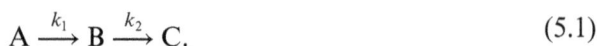

While a sequence of completely irreversible steps is uncommon in chemistry, such mechanisms are commonplace in processes such as radioactive decay. For example

$$^{238}_{92}U \xrightarrow{-\alpha} {}^{234}_{90}Th \xrightarrow{-\beta} {}^{234}_{91}Pa. \tag{5.2}$$

A mechanism of this type represents one of the few kinetic schemes for which it is fairly straightforward to solve the rate equations analytically, and we will explore the kinetics in some detail for this reason.

It is straightforward to identify a number of initial conditions for the reactant and product concentrations. Assuming that we begin with the only reactant present, at concentration $[A]_0$, we can immediately conclude that at time $t = 0$ the concentrations of A, B, and C are $[A] = [A]_0$, $[B] = [C] = 0$. Since the reaction transforms A into B and C in a stepwise manner and in a 1:1 ratio, we also have the condition that at all times $[A] + [B] + [C] = [A]_0$. We are now ready to set up the rate equations describing the changes in $[A]$, $[B]$, and $[C]$ with time, and to solve them to determine the concentrations of A, B, and C as a function of time.

The rate equations are[1]:

$$\frac{d[A]}{dt} = -k_1[A] \tag{5.3}$$

$$\frac{d[B]}{dt} = k_1[A] - k_2[B] \tag{5.4}$$

$$\frac{d[C]}{dt} = k_2[C]. \tag{5.5}$$

Integrating the first rate equation (see section 2.4) gives

$$[A] = [A]_0 e^{-k_1 t}. \tag{5.6}$$

Substituting this into the second rate equation gives another first-order differential equation,

$$\frac{d[B]}{dt} + k_2[B] = k_1[A]_0 e^{-k_1 t}, \tag{5.7}$$

---

[1] The concentration of A changes only as a result of the first reaction step, and the rate of change of $[A]$ is therefore $d[A]/dt = -k_1[A]$. The concentration of B changes as it is formed in the first step and used up in the second step, giving an overall rate of change $d[B]/dt = k_1[A] - k_2[B]$. The concentration of C changes only as a result of being formed in the second step, giving an overall rate of change $d[C]/dt = k_2[B]$.

which has the solution

$$[B] = \frac{k_1}{k_2 - k_1}(e^{-k_1 t} - e^{-k_2 t})[A]_0. \tag{5.8}$$

Finally, since $[C] = [A]_0 - [B] - [A]$, we find

$$[C] = \left(1 + \frac{k_1 e^{-k_2 t} - k_2 e^{-k_1 t}}{k_2 - k_1}\right)[A]_0. \tag{5.9}$$

These equations allow us to predict the way in which the concentrations of reactant A, intermediate B, and product C will change with time for any values of the rate constants k_1 and k_2. Considering the extreme cases, when either $k_1 \gg k_2$ or $k_2 \gg k_1$, proves to be rather instructive in allowing us to introduce a number of additional features of complex reactions. These two cases are shown in figure 5.1.

5.1.1 Case 1: $k_1 \gg k_2$

When the rate of the first step in the reaction sequence is much faster than that of the second, all of the A initially present is rapidly converted into B, which is then slowly used up to form C. We can treat k_2 as negligible in comparison with k_1 in the denominator of equation (5.9), and the expression simplifies to

$$[C] = (1 - e^{-k_2 t})[A]_0. \tag{5.10}$$

We see that the rate of production of C, and therefore the overall rate of the two-step reaction, becomes independent of k_1 and only depends on the rate constant k_2. We say that the second, 'slow' step is the *rate determining step* in the mechanism.

5.1.2 Case 2: $k_2 \gg k_1$

When we reverse the situation so that the second step in the mechanism becomes much faster than the first, the intermediate B is consumed as soon as it is produced. We will see later that this has significant consequences when we consider reactive

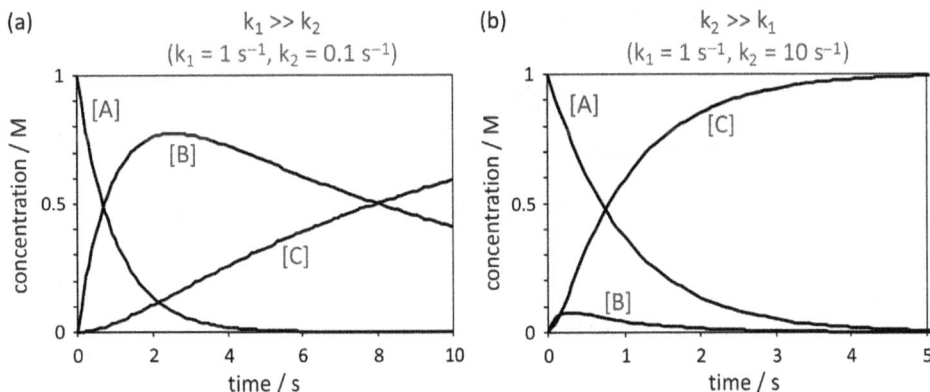

Figure 5.1. Evolution of the concentrations of A, B, and C for the reaction $A \xrightarrow{k_1} B \xrightarrow{k_2} C$ for the cases when $k_1 \gg k_2$ and $k_2 \gg k_1$.

intermediates in chapter 6. We can now treat k_1 as negligible in relation to k_2 in the denominator of equation (5.9), and the expression now simplifies to

$$[C] = (1 - e^{-k_1 t})[A]_0. \tag{5.11}$$

The overall rate of production of C now depends only on the rate constant k_1, and the first step is now rate determining.

5.2 Pre-equilibria

In our second example, we will make the mechanism only slightly more complicated than the sequential reaction mechanism above by making the first step reversible.

$$A + B \underset{k_{-1}}{\overset{k_1}{\rightleftharpoons}} C \overset{k_2}{\longrightarrow} D. \tag{5.12}$$

The rate equations are now:

$$\frac{d[A]}{dt} = \frac{d[B]}{dt} = -k_1[A][B] + k_{-1}[C]$$

$$\frac{d[C]}{dt} = k_1[A][B] - k_{-1}[C] - k_2[C] \tag{5.13}$$

$$\frac{d[D]}{dt} = k_2[C].$$

Despite the fact that the reaction mechanism is only a little more complicated than for the sequential reaction scheme explored previously, these equations cannot be solved analytically, and in general would have to be integrated numerically to obtain an accurate solution. However, the situation simplifies considerably for the special case in which $k_{-1} \gg k_2$. In this case, an equilibrium is reached between the reactants A and B and the intermediate C, and the equilibrium is only perturbed very slightly by C 'leaking away' very slowly to form the product D.

If we assume that we can neglect this small perturbation of the equilibrium, then, once equilibrium is reached, the rates of the forward and reverse reactions must be equal.

$$k_1[A][B] = k_{-1}[C]. \tag{5.14}$$

Rearranging the above equation, we find

$$\frac{k_1}{k_{-1}} = \frac{[C]}{[A][B]}. \tag{5.15}$$

We recognise the right hand side of this equation as the equilibrium constant, K, and see that as well as being expressible in terms of the product and reactant concentrations, the equilibrium constant can also be expressed as the ratio of the rate constants for the forward and reverse reactions, i.e. $K = k_1/k_{-1}$.

The rate of the overall reaction is simply the rate of formation of the product D, so, using the above result, we can write

$$\frac{d[\text{D}]}{dt} = k_2[\text{C}] = k_2 K[\text{A}][\text{B}].$$ (5.16)

The reaction is therefore first order in each of the reactant concentrations [A] and [B], and second order overall, with an effective rate constant $k_{\text{eff}} = k_2 K$. It is worth noting that because this rate law has been derived based on the assumption of a pre-equilibrium between A, B, and C, it will not be accurate in the very early stages of the reaction, before the equilibrium has been established.

5.3 Moving on to more complicated mechanisms

At this point, we have considered two simple reaction mechanisms for which the rate equations can be solved exactly—or nearly exactly in the case of the pre-equilibrium. Unfortunately, aside from these two simple examples, the rate equations for virtually all complex reaction mechanisms generally comprise a complicated system of coupled differential equations that cannot be solved analytically. In state-of-the-art kinetic modelling studies, sophisticated software is used to obtain numerical solutions to the rate equations in order to determine the time-varying concentrations of all species involved in a reaction sequence. This is an extremely valuable approach when accurate information is required for comparison with experiment or for predictive purposes. However, it often does not provide the same insights that can be obtained from an analytical rate law of the type derived in the two preceding examples. Luckily, by making a few simple assumptions about the nature of reactive intermediates within a reaction sequence, very good approximate solutions to the rate equations may be derived, which do provide such insights. This approach will be the topic of chapter 6

An Introduction to Chemical Kinetics

Claire Vallance

Chapter 6

Using the steady-state approximation to derive rate laws for complex reactions

In this chapter, we introduce the steady-state approximation, which will allow us to derive approximate rate laws even for reactions with highly complex reaction mechanisms. The steady-state approximation relies on assumptions relating to reactive intermediates involved in the reaction of interest. An intermediate is a species that is formed during the reaction but is completely transformed into another species during one or more elementary steps, and so does not appear in the overall reaction equation.

Almost by definition, a reactive intermediate in a reaction mechanism is used up virtually as soon as it is formed, and therefore its concentration remains very low and essentially constant throughout the course of the reaction. This is true at all times apart from at the very start of the reaction, when the concentration [R] of the reactive intermediate must necessarily build up from zero to some small non-zero value, and at the very end of the reaction (at least in the case of a reaction that goes to completion), when [R] must return to zero. During the period of time when [R] is essentially constant, we find that because $d[R]/dt$ is so much smaller than the rates of change of the reactant and product concentrations, it is a good approximation to set $d[R]/dt = 0$. This is known as the *steady state approximation*.

Steady-state approximation: if a reactive intermediate R is present at constant concentration throughout (most of) the course of the reaction, then we can set $d[R]/dt = 0$ in the rate equations.

As we shall see, applying the steady state approximation has the effect of converting a mathematically intractable set of coupled differential equations into a system of simultaneous algebraic equations, one equation for each species involved in the reaction. The algebraic equations may be solved to find the concentrations of the reactive intermediates, and these may then be substituted back into the equation for the rate of change of the product in order to obtain an expression for the overall rate law.

6.1 The steady-state approximation: a first example

As a simple example, let us look again at the reaction scheme we considered in section 5.2,

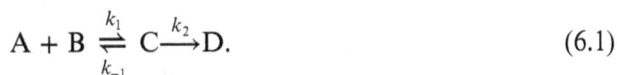

$$A + B \underset{k_{-1}}{\overset{k_1}{\rightleftharpoons}} C \overset{k_2}{\longrightarrow} D. \tag{6.1}$$

This time, instead of assuming a pre-equilibrium, we will investigate the case in which $k_2 \gg k_{-1}$, so that C is now a reactive intermediate and there is no stable equilibrium between A, B and C. We can apply the steady state approximation (SSA) to C, to obtain

$$\frac{d[C]}{dt} = 0 = k_1[A][B] - k_{-1}[C] - k_2[C]. \tag{6.2}$$

This expression may be rearranged to give [C] in terms of the reactant concentrations [A] and [B].

$$[C] = \frac{k_1}{k_{-1} + k_2}[A][B]. \tag{6.3}$$

The overall reaction rate is the rate of formation of the product, D.

$$\frac{d[D]}{dt} = k_2[C] = \frac{k_1 k_2}{k_{-1} + k_2}[A][B]. \tag{6.4}$$

In contrast to the previous treatment of this reaction, in which we assumed a pre-equilibrium and obtained a rate law valid only in the case when $k_{-1} \gg k_2$, the above rate law is valid so long as the concentration of C has reached a steady state. We can consider a couple of limiting cases.

In the limiting case where k_{-1} is much smaller than k_2, we can neglect k_{-1} in the denominator and the rate law simplifies to give an overall rate equal to k_1 [A] [B], i.e. the rate of the overall reaction is the same as the rate of the first elementary step in the reaction mechanism. This is not all that surprising. If k_2 is much larger than k_1 and k_{-1}, then as soon as the A + B $\longrightarrow$ C step has occurred, C is immediately converted into products, and there is virtually no chance for the reverse reaction, C $\longrightarrow$ A + B, to occur. The initial elementary step is rate determining, and therefore dominates the kinetics.

In the limiting case where $k_{-1} \gg k_2$, we can neglect k_2 in the denominator, and the rate law simplifies to give a rate equal to $(k_1 k_2/k_{-1})$[A][B]. Reassuringly, this is the same result as we obtained in our previous treatment by assuming a pre-equilibrium, with $k_1/k_{-1} = K$.

6.2 The steady-state approximation: a general approach

We will consider many more examples of applying the steady-state approximation over the the course of the next two chapters, and will apply the same general approach in each case.

1. Write down a steady-state equation for each reactive intermediate, R_i, by setting $d[R_i]/dt = 0$.

2. Solve the resulting set of simultaneous equations to obtain expressions for the concentrations of each intermediate in terms of the reactant and product concentrations. It is worth looking carefully at the equations in order to identify the simplest route to solving them. For example:

   - If one of the equations contains only one reactive intermediate, it may simply be rearranged to give the concentration of that intermediate in terms of reactant and product concentrations. The resulting expression can often be substituted into other equations to obtain the corresponding expressions for other reactive intermediates.
   - If the equations depend on more than one reactive intermediate, and share terms, look for sums or differences of the equations that will simplify matters. Often a steady-state problem that initially appears extremely complicated becomes trivial when you simply add together two of the steady state equations.

3. Write down an expression for the overall rate (usually the rate of change of one of the products). This will generally involve the concentrations of one or more reactive intermediates.

4. Substitute your expressions for the reactive intermediate concentrations into the overall rate equation in order to eliminate reactive intermediates from the equation. The result should be an overall rate equation that depends only on the reactant and product concentrations. Concentrations of reactive intermediates must not appear in the final rate law. If they do, then you have not finished solving the steady-state equations.

Armed with a general 'recipe' for applying the steady-state approximation, we will now use the approach to determine the rate law for a number of commonly encountered reaction types.

6.3 'Unimolecular' reactions: the Lindemann–Hinshelwood mechanism

A number of gas-phase reactions follow first order kinetics and apparently only involve one chemical species. Examples include the structural isomerisation of cyclopropane to propene, and the decomposition of azomethane ($CH_2N_2CH_3 \longrightarrow C_2H_6 + N_2$, with experimentally determined rate law $\frac{d[CH_2N_2CH_3]}{dt} = k[CH_3N_2CH_3]$). The mechanism by which the reactant molecules acquire enough energy to react remained a puzzle for some time, particularly since the rate law seemed to rule out a bimolecular step. The puzzle was solved by Lindemann in 1922, when he proposed the following mechanism for 'thermal' unimolecular reactions[1].

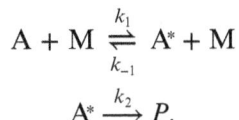

$$A + M \underset{k_{-1}}{\overset{k_1}{\rightleftharpoons}} A^* + M$$

$$A^* \xrightarrow{k_2} P.$$

The reactant, A, acquires enough energy to react by colliding with another molecule, M (note that in many cases M will actually be another A molecule). The excited reactant A* then undergoes unimolecular reaction to form the products, P. To determine the overall rate law arising from this mechanism, we can apply the steady-state approximation to the excited state (reactive intermediate) A*.

$$\frac{d[A^*]}{dt} = 0 = k_1[A][M] - k_{-1}[A^*][M] - k_2[A^*]. \tag{6.5}$$

Rearranging this expression yields the concentration of the reactive intermediate, [A*].

$$[A^*] = \frac{k_1[A][M]}{k_{-1}[M] + k_2}. \tag{6.6}$$

The overall rate of reaction is then

$$\frac{d[P]}{dt} = k_2[A^*] = \frac{k_1 k_2[A][M]}{k_{-1}[M] + k_2}. \tag{6.7}$$

At first sight, this does not look very much like the first order rate law we were expecting. However, consider the behaviour of this rate law in the limits of high and low pressure.

6.3.1 High pressure

At high pressure there are many collisions, and collisional de-excitation of A* is therefore much more likely than an unimolecular reaction of A* to form products, i.e. $k_{-1}[A^*][M] \gg k_2[A^*]$. In this limit, we can neglect the k_2 term in the denominator of equation (6.7), and the rate law simplifies to

$$\frac{d[P]}{dt} = \frac{k_1 k_2}{k_{-1}}[A]. \tag{6.8}$$

This is a first order rate law, with a first order ('unimolecular') rate constant $k_{uni} = k_1 k_2/k_{-1}$. The proposed mechanism therefore explains the observed first-order kinetics when the reaction is carried out at sufficiently high pressures, under conditions for which the unimolecular step in the mechanism becomes rate determining.

---

[1] Unimolecular reactions, and indeed many other types of reaction, may also be initiated photochemically by absorption of a photon.

6.3.2 Low pressure

At low pressures, there are few collisions, and A* will generally undergo the unimolecular reaction before it undergoes collisional de-excitation, i.e. $k_2[A^*] \gg k_{-1}[A^*][M]$. In this case, we can neglect the $k_{-1}[M]$ term in the denominator of equation (6.7), and the rate law simplifies to

$$\frac{d[P]}{dt} = k_1[A][B].\tag{6.9}$$

Under low-pressure conditions, the kinetics are therefore second order. Formation of the excited species A*, a bimolecular process, is now the rate determining step.

6.3.3 Testing the rate law against experimental data

While the proposed mechanism has explained the observation of first-order kinetics when the reaction is carried out at sufficiently high pressures, any proposed rate law should show quantitative agreement with experimental data across all pressures in order to inspire confidence in the proposed reaction mechanism. In the present case, we can rewrite the rate law (equation (6.7)) in the following form:

$$\frac{d[P]}{dt} = k_{\text{eff}}[A]\tag{6.10}$$

with the effective rate constant

$$k_{\text{eff}} = \frac{k_1 k_2[M]}{k_{-1}[M] + k_2}.\tag{6.11}$$

The rate constant k_{eff} is a first-order rate constant, but it depends on the concentration of the collision partner M, and only truly becomes constant at sufficiently high pressure, when it reduces to the value $k_{\text{uni}} = k_1 k_2 / k_{-1}$, as shown previously.

If experimental measurements of the rate constant k_{eff} are made as a function of pressure (equivalent to [M]), then the Lindemann–Hinshelwood mechanism may be tested. Taking the reciprocal[2] of equation (6.11) gives

$$\frac{1}{k_{\text{eff}}} = \frac{k_{-1}}{k_1 k_2} + \frac{1}{k_1[M]}.\tag{6.12}$$

A plot of $1/k_{\text{eff}}$ against $1/[M]$ should therefore be linear, with an intercept of $k_{-1}/(k_1 k_2)$ and a slope of $1/k_1$. An example of such a plot is shown in figure 6.1.

Usually there is a reasonable fit between theory and experiment at low pressure, but a pronounced deviation at high pressure, with experimental values of k_{eff} being larger than the values predicted by the Lindemann–Hinshelwood mechanism. The proposed mechanism therefore correctly predicts some, but not all, features of the reaction. It turns out that while the general idea of a collisional activation process is correct, the

---

[2] It is useful to note that many rate laws can be transformed into a form suitable for testing via a straight-line plot of experimental data by taking the reciprocal.

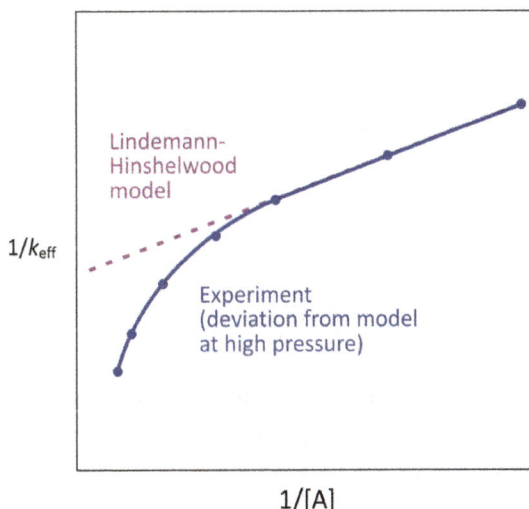

Figure 6.1. Comparison of experimental data with the predictions of the Lindemann–Hinshelwood mechanism.

true mechanism of 'unimolecular' reactions is slightly more involved. The principal failing of the Lindemann–Hinshelwood mechanism is that it assumes that *any* excited reactant A* will undergo an unimolecular reaction to produce products. In practice, however, excitation is generally required in a degree of freedom that is coupled to the reaction coordinate in some way. For example, vibrational excitation may be required in a bond that breaks during the reaction. More sophisticated theories of unimolecular reactions have been developed which take this and other factors into account, providing much better agreement with experiment[3].

6.4 Third-order reactions

A number of reactions are found experimentally to have third-order kinetics. An example is the oxidation of NO, for which the overall reaction equation and experimentally-determined rate law are given below.

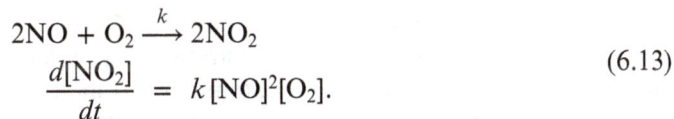

$$2NO + O_2 \xrightarrow{k} 2NO_2$$
$$\frac{d[NO_2]}{dt} = k\,[NO]^2[O_2].$$
(6.13)

One possibility for the reaction mechanism is a single elementary step involving a three-body collision, i.e. a true termolecular reaction. However, such collisions are exceedingly rare, and certainly too unlikely to explain the observed rate at which this reaction proceeds. An added complication is that the reaction rate is found to decrease with increasing temperature, a certain indication of a complex mechanism.

---

[3] More sophisticated models include transition state theory, RRK theory, and RRKM theory, described in more advanced texts on reaction kinetics.

An alternative mechanism that leads to the same rate law consists of two elementary steps involving a pre-equilibrium.

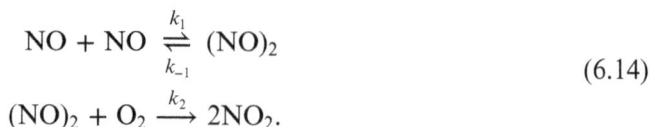

$$NO + NO \underset{k_{-1}}{\overset{k_1}{\rightleftharpoons}} (NO)_2$$

$$(NO)_2 + O_2 \xrightarrow{k_2} 2NO_2. \tag{6.14}$$

The overall rate, expressed as the rate of formation of products, is

$$\frac{1}{2}\frac{d[NO_2]}{dt} = k_2[(NO)_2][O_2]. \tag{6.15}$$

However, under the conditions of a pre-equilibrium (see section 5.2), we have

$$K = \frac{[(NO)_2]}{[NO]^2} \tag{6.16}$$

where K is the equilibrium constant for the pre-equilibrium between NO and $(NO)_2$, and therefore

$$[(NO)_2] = K[NO]^2. \tag{6.17}$$

Substituting this result into our expression for the overall rate gives

$$\frac{1}{2}\frac{d[NO_2]}{dt} = k_2 K[NO]^2[O_2] \tag{6.18}$$

which is a third-order rate law, as required.

A very common situation in which third-order kinetics are observed involves reactions in which two reactants combine to form a single product. Such reactions require a so-called 'third body' to take away some of the excess energy from the reaction product. An example is the formation of ozone.

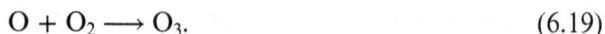

$$O + O_2 \longrightarrow O_3. \tag{6.19}$$

If the mechanism consisted of a single elementary step, as written, this reaction would barely occur. To understand the reason for this, we need to turn to some basic classical mechanics, namely the fact that in any collision, energy and momentum must both be conserved. To demonstrate the problem, consider the somewhat contrived situation in which the O atom and O_2 molecule initially have equal and opposite momenta, and collide head-on to react. The following arguments apply equally well to any other situation, but are clearest to see in this simple case[4]. Since

---

[4] The momentum of the centre of mass of the system must be conserved in any collision, and so any kinetic energy associated with this motion is not available for the reaction. This means that only the relative motion of the reactants is important in the reaction. This is precisely the motion that is considered in the example. Any reactive collision can be transformed into the so-called 'centre-of-mass' frame by subtracting the momentum of the centre of mass from the momenta of each reactant, without changing the physics of the collision. The conclusions reached in the example are therefore completely general, and apply equally well in systems for which the total momentum is non-zero.

total momentum must be conserved, and initially the total momentum is zero (because the momenta of the O and O_2 exactly cancel each other out), the final momentum of the O_3 product must also be zero i.e. the product molecule must be stationary. Now consider the conservation of energy. Since we are forming a bond, the reaction is exothermic, so by the conservation of energy, the total kinetic energy possessed by the O_3 product must be the sum of the reaction exothermicity and the kinetic energies of the reactants. We have already determined that conservation of momentum requires the newly formed O_3 molecule to be stationary, so all of this kinetic energy must be channelled into the vibrational and/or rotational motion of the molecule, and not into translational motion. Highly vibrationally excited molecules are extremely unstable, and under the circumstances described the O_3 will very quickly dissociate back into reactants. The only way for the O_3 to survive is for it to transfer some of its vibrational energy to another molecule, M, in a collision. The molecule M is known as a *third body*. The energy may end up as internal excitation (rotation or vibration) of M, or simply as kinetic energy as the two molecules fly away from each other after the collision. The actual mechanism is therefore

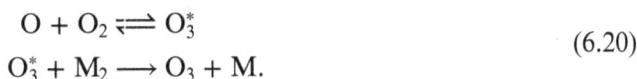

$$O + O_2 \rightleftharpoons O_3^*$$
$$O_3^* + M_2 \longrightarrow O_3 + M. \tag{6.20}$$

The overall reaction is usually written as $O + O_2 + M \longrightarrow O_3 + M$.

Note that *a third body is only required for reactions in which a single product is formed from two or more reactants*, since this is the only time that the conservation of momentum forces a large amount of energy to be released into the internal excitation of the product. If two products are formed, they can both carry away almost arbitrary amounts of energy as translational kinetic energy, while still conserving the total momentum.

6.5 Enzyme reactions and the Michaelis–Menten mechanism

An enzyme is a protein that catalyses a chemical reaction by lowering the activation energy. Each enzyme has an active site that is carefully designed by nature to bind a particular reactant molecule, known as the substrate. An example of a substrate bound at the active site of an enzyme is shown in figure 6.2. The activation energy of the reaction for the enzyme-bound substrate is lower than for the free substrate molecule, due to the fact that the interactions involved in binding shift the substrate geometry closer to that of the transition state for the reaction. Once the reaction has occurred, the product molecules are released from the enzyme.

Enzyme-catalysed reactions occur millions of times faster than the corresponding uncatalysed reactions. Virtually every chemical reaction in biology requires an enzyme in order to occur at a significant rate, and each enzyme is specific to a particular reaction. Many drugs work by binding to a carefully targeted enzyme in place of the normal substrate molecule, thereby inhibiting enzyme activity and slowing the reaction rate. Enzyme kinetics is an extremely important and complex

Binding site binds
and orients substrate
molecule

Catalytic site distorts
substrate structure to
lower activation barrier.

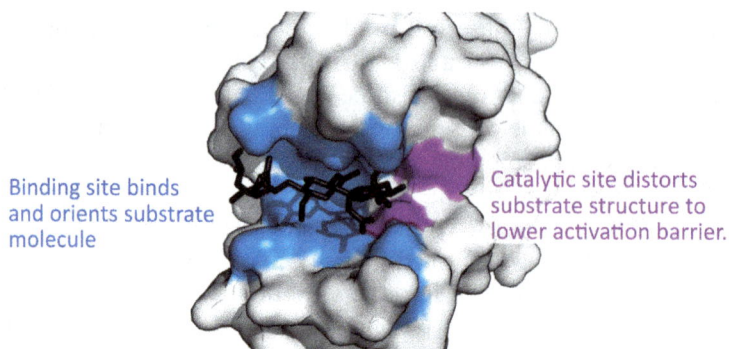

Figure 6.2. An enzyme-substrate complex (peptidoglycan bound to lysozyme). Adapted from a figure by Thomas Shafee under Creative Commons license.

field, but the basic kinetics of a simple enzyme catalysis process may be modelled quite simply. We begin with a brief overview of the experimentally observed kinetics.

In an enzyme-catalysed reaction, a substrate S is converted to products P in a reaction that is catalysed by an enzyme E. For many such reactions, the rate is found experimentally to follow the Michaelis–Menten equation

$$\frac{d[P]}{dt} = \frac{\nu_{max}[S]}{K_M + [S]}. \tag{6.21}$$

The constant K_M is called the Michaelis constant, and ν_{max} is the maximum reaction rate, which is found to be linearly proportional to the total enzyme concentration.

$$\nu_{max} = k_{cat}[E]_0. \tag{6.22}$$

The constant of proportionality, k_{cat} is known as the *turnover number*, and represents the maximum number of molecules of substrate that each enzyme molecule can convert into products (or 'turn over') per second. We shall see later that this maximum rate occurs when the substrate is present in large excess.

Any kinetic model for enzyme catalysis must explain the fact that the rate depends on the enzyme concentration [E], even though there is no net change in its concentration over the course of the reaction. The simplest trial mechanism involves the formation of a bound enzyme–substrate complex ES, followed by conversion of the complex into the products and free enzyme (which may then go on to catalyse further reactions).

$$E + S \underset{k_{-1}}{\overset{k_1}{\rightleftharpoons}} ES \overset{k_2}{\longrightarrow} P + E. \tag{6.23}$$

This mechanism represents a slightly different example of applying the steady state approximation from those we have encountered so far. In the previous cases, the species identified as reactive intermediates had concentrations much lower than those of the reactants. In the case of an enzyme-catalysed reaction, the concentration of the reactive intermediate ES is not much less than the free enzyme concentration,

[E]. However, because the enzyme is regenerated in the second step of the mechanism, both [E] and [ES] change much more slowly than [S] and [P], and so the steady-state approximation is valid. Applying the steady-state approximation to [ES], we have

$$\frac{d[ES]}{dt} = 0 = k_1[E][S] - k_{-1}[ES] - k_2[ES]. \tag{6.24}$$

Solving for the concentration of ES, we obtain

$$[ES] = \frac{k_1[E][S]}{k_{-1} + k_2}. \tag{6.25}$$

If the total enzyme concentration is $[E]_0$, and the enzyme is present either as free enzyme, E, or enzyme–substrate complex, ES, then the amount of free enzyme must be $[E] = [E]_0 - [ES]$. Substituting this into the above equation gives

$$[ES] = \frac{k_1([E]_0 - [ES])[S]}{k_{-1} + k_2} \tag{6.26}$$

which rearranges to give

$$[ES] = \frac{k_1[E]_0[S]}{k_{-1} + k_2 + k_1[S]}. \tag{6.27}$$

The overall rate of reaction is then found from the rate of formation of product, P.

$$\begin{aligned} \frac{d[P]}{dt} &= k_2[ES] \\ &= \frac{k_2 k_1[E]_0[S]}{k_{-1} + k_2 + k_1[S]} \\ &= \frac{k_2[S][E]_0}{K_M + [S]} \\ &= k[E]_0 \end{aligned} \tag{6.28}$$

where

$$k = \frac{k_2[S]}{K_M + [S]} \tag{6.29}$$

and the Michaelis constant K_M is given by

$$K_M = \frac{k_2 + k_{-1}}{k_1}. \tag{6.30}$$

While the definitions of k and K_M above may seem fairly arbitrary, we have chosen these particular combinations of rate constants in order to finish with a rate equation in the same form as the experimentally-derived Michaelis–Menten equation. The two equations (6.28) and (6.21) now agree if we define $k_2 = k_{cat}$.

We can see from the above treatment that the rate of enzyme-catalysed reaction depends linearly on the enzyme concentration, [E], but in a more complicated way

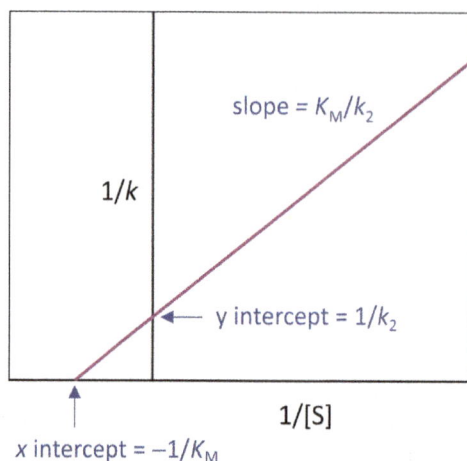

Figure 6.3. A Lineweaver–Burk plot of $1/k$ against $1/[S]$ for an enzyme-catalysed reaction. The plot has a slope of K_M/k_2 and an intercept of $1/k_2$, allowing K_M and k_2 to be determined.

on the substrate concentration, $[S]$. The dependence on $[S]$ simplifies under certain conditions.

1. When $[S] \gg K_M$, then the rate becomes

$$\frac{d[P]}{dt} = k_2 [E]_0 = k_{cat} [E]_0 \tag{6.31}$$

and the overall rate is independent of the substrate concentration. There is so much substrate present that only a tiny fraction is bound up in the enzyme–substrate complex and used up in the reaction, and the concentration of the free substrate remains essentially constant as the reaction proceeds. The enzyme is saturated with substrate, and the reaction rate reaches a maximum.

2. When $[S] \ll K_M$, the reaction rate becomes

$$\frac{d[P]}{dt} = \frac{k_2}{K_M}[E]_0[S] \tag{6.32}$$

and the rate is first order in both $[E]_0$ and $[S]$.

To determine k_2 and K_M from experimental rate data, we invert the expression for k (equation (6.29)) to obtain a new form of the equation that is expected to generate a straight line plot known as a Lineweaver–Burk plot (see figure 6.3 for an example)

$$\frac{1}{k} = \frac{K_M}{k_2[S]} + \frac{1}{k_2}. \tag{6.33}$$

A plot of $1/k$ against $1/[S]$ has a slope of K_M/k_2, a 'y' intercept of $1/k_2$, and an 'x' intercept of $1/K_M$. We can therefore find K_M and k_2 directly from the x and y

intercepts, or from one intercept and the slope. Usually, the initial rates method is used to measure k, in order to preclude any complications that may arise from secondary reactions of the products.

Having looked at a few relatively straightforward examples of chemical reaction mechanisms and their associated kinetics, we are ready to study some more complicated chemical systems.

An Introduction to Chemical Kinetics

Claire Vallance

Chapter 7

Chain reactions and explosions

7.1 Chain reactions

Chain reactions are complex reactions that involve *chain carriers*, the name given to reactive intermediates that react to produce further reactive intermediates. Some examples of chain reactions are:

- combustion of a fuel gas;
- development of rancidity in fats;
- the polymerase chain reaction (PCR) used to amplify DNA samples for analysis;
- many polymerisation reactions, e.g. polymerisation of ethene to polyethene;
- nuclear fission caused by neutron bombardment.

The elementary steps in a chain reaction may be classified into *initiation*, *propagation*, *inhibition*, and *termination* steps. As an example, consider the infamous reaction by which chlorofluorocarbons (CFCs) destroy ozone:

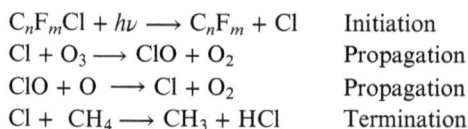

$$
\begin{aligned}
C_nF_mCl + h\nu &\longrightarrow C_nF_m + Cl && \text{Initiation} \\
Cl + O_3 &\longrightarrow ClO + O_2 && \text{Propagation} \\
ClO + O &\longrightarrow Cl + O_2 && \text{Propagation} \\
Cl + CH_4 &\longrightarrow CH_3 + HCl && \text{Termination}
\end{aligned}
$$

The various steps in the mechanism are classified as follows:

1. *Initiation step*
   The reaction is initiated either thermally or photochemically. The first reactive intermediates/chain carriers (in this case a Cl radical) are produced.

2. *Propagation step*
   The reaction of a radical leads to the formation of another radical, i.e. another reactive intermediate. In the first propagation step above, Cl reacts to form ClO; in the second step ClO reacts to form Cl.

doi:10.1088/978-1-6817-4664-7ch7
© Morgan & Claypool Publishers 2017

3. *Termination step*
   Chain carriers are deactivated. Often this occurs through radical–radical recombination, reactions with walls, or reactions with another molecule to create an inactive product. Note that termination products may go on to be involved in other reactions, but are not involved further in the chain reaction of interest.

Some chain reactions also involve inhibition steps, in which product molecules are destroyed. Inhibition steps are sometimes also referred to as *retardation* or *de-propagation* steps.

The *chain length* in a chain reaction is defined as the number of propagation steps per initiation step, or alternatively as the rate of propagation divided by the rate of initiation. Chain lengths can be very long: in the above example a single Cl radical can destroy around 10^6 molecules of ozone

$$\text{chain length}, n = \frac{\text{rate of propagation}}{\text{rate of initiation}} = \frac{\text{rate of propagation}}{\text{rate of termination}}. \tag{7.1}$$

The reaction of Cl atoms with ozone is an example of a *cyclic chain reaction*. Atomic chlorine acts as a catalyst and is continuously regenerated until it is removed by a termination step. It is also possible to have *non-cyclic chain reactions*, involving many reactive species and elementary steps. Non-cyclic chain reactions can have extremely complicated kinetic mechanisms.

Chain reactions in which each propagation step produces only one reactive intermediate are called *linear chain reactions*. *Branched chain reactions* are also possible, in which a chain carrier reacts to form more than one chain carrier in a single elementary step. We will look at some examples of both linear and branched chain reactions in the following.

7.2 Linear chain reactions

The reaction between H_2 and Br_2 has become the 'benchmark' system for illustrating the kinetics of linear chain reactions, and we will use this reaction as our main example. While the reaction is not the most up-to-date of examples, having first been studied around 100 years ago, it provides a very clear illustration of all of the key features of chain reactions. Once we have studied the hydrogen-bromine reaction, we can also investigate the reactions of H_2 with Cl_2 and I_2. While we might expect the three reactions to have identical mechanisms, we in fact find them to be markedly different, providing an opportunity to investigate a variety of factors that determine the most favoured (i.e. fastest) mechanism.

7.2.1 The hydrogen-bromine reaction

As noted above, the kinetics of the reaction between H_2 and Br_2 were determined experimentally by Bodenstein and Lind around 100 years ago[1]. The overall reaction equation is

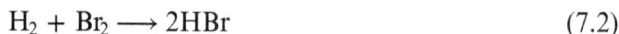

$$H_2 + Br_2 \longrightarrow 2HBr \tag{7.2}$$

and Bodenstein initially determined the rate law to be

$$\frac{d[HBr]}{dt} = k\,[H_2][Br_2]^{1/2}. \tag{7.3}$$

The measured order of 1/2 with respect to Br_2 indicated that the reaction proceeds via a complex reaction mechanism, rather than a simple bimolecular collision. Further investigation showed that this rate law in fact only holds for the early stages of the reaction, and that the true rate law takes the form

$$\frac{d[HBr]}{dt} = \frac{k\,[H_2][Br_2]^{1/2}}{1 + k'[HBr]/[Br_2]}. \tag{7.4}$$

Any proposed mechanism for the reaction must agree with both of these observations.

The first step in any chain reaction is the initiation step. The reaction between H_2 and Br_2 can be initiated by either thermally-induced or photon-induced dissociation of Br_2

$$Br_2 + M \longrightarrow Br + Br + M \tag{7.5}$$

or

$$Br_2 + h\nu \longrightarrow Br + Br. \tag{7.6}$$

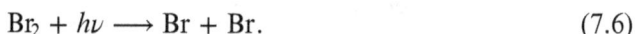

We will concentrate on the thermal mechanism for the purposes of deriving a rate law for the overall reaction, but the steps following the initiation step are the same for both cases. The currently accepted mechanism is:

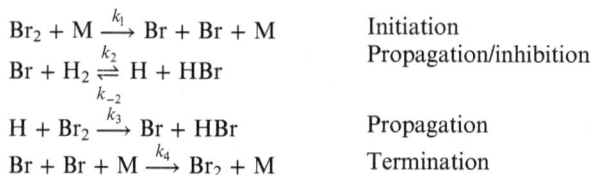

$$
\begin{array}{ll}
Br_2 + M \xrightarrow{k_1} Br + Br + M & \text{Initiation} \\
Br + H_2 \underset{k_{-2}}{\overset{k_2}{\rightleftharpoons}} H + HBr & \text{Propagation/inhibition} \\
H + Br_2 \xrightarrow{k_3} Br + HBr & \text{Propagation} \\
Br + Br + M \xrightarrow{k_4} Br_2 + M & \text{Termination}
\end{array}
$$

In the second step, because the H–H bond is stronger than the H–Br bond, the reverse (inhibition) reaction becomes possible once an appreciable amount of HBr has built up. The reaction chain contains two radical chain carriers, H and Br. In order to arrive at an overall rate law for the reaction, we apply the steady-state approximation to the two chain carriers.

---

[1] M Bodenstein and C Lind 1906 *Z. Phys. Chem.*, **57** 168.

$$\frac{d[\mathrm{H}]}{dt} = 0 = k_2[\mathrm{Br_2}][\mathrm{H_2}] - k_{-2}[\mathrm{H}][\mathrm{Br}] - k_3[\mathrm{H}][\mathrm{Br_2}]$$

$$\frac{d[\mathrm{Br}]}{dt} = 0 = 2k_1[\mathrm{Br_2}][\mathrm{M}] - k_2[\mathrm{Br}][\mathrm{H_2}] + k_{-2}[\mathrm{H}][\mathrm{HBr}] + k_3[\mathrm{H}][\mathrm{Br_2}] \tag{7.7}$$

$$- 2k_4[\mathrm{Br}]^2[\mathrm{M}].$$

We can solve these two equations to obtain expressions for the concentrations of H and Br in terms of the reactant and product concentrations and the various rate constants. This is a good example of a case in which we can simplify the required algebra considerably through careful inspection. The two equations each depend on both chain carrier concentrations, and also share terms. We can simplify the process of solving the equations by adding them together to give:

$$0 = 2k_1[\mathrm{Br_2}][\mathrm{M}] - 2k_4[\mathrm{Br}]^2[\mathrm{M}]. \tag{7.8}$$

This expression can be rearranged to give the steady-state concentration of atomic bromine.

$$[\mathrm{Br}] = \left(\frac{k_1[\mathrm{Br_2}]}{k_4} \right)^{1/2}. \tag{7.9}$$

Note that equation (7.8) implies that the rate of initiation is the same as the rate of termination, as expected under steady-state conditions. This provides a good check that we have made no mistakes up to this point. Equation (7.8) also leads to a considerable simplification of the steady-state equation describing the rate of change of [Br], which now becomes

$$0 = -k_2[\mathrm{Br}][\mathrm{H_2}] + k_{-2}[\mathrm{H}][\mathrm{HBr}] + k_3[\mathrm{H}][\mathrm{Br_2}] \tag{7.10}$$

and may be rearranged to give an expression for the steady-state concentration of H atoms.

$$[\mathrm{H}] = \frac{k_2[\mathrm{Br}][\mathrm{H_2}]}{k_2[\mathrm{HBr}] + k_3[\mathrm{Br_2}]}$$

$$= \frac{k_2[\mathrm{H_2}]}{k_2[\mathrm{HBr}] + k_3[\mathrm{Br_2}]} \left(\frac{k_1[\mathrm{Br_2}]}{k_4} \right)^{1/2} \tag{7.11}$$

where in the second step above we have substituted in our previously determined expression (equation (7.9)) for the concentration of Br. Now that we have determined the concentrations of the reactive intermediates, we are ready to determine the overall reaction rate. This is found from the rate of change of the concentration of the HBr product. From the reaction mechanism, we have

$$\frac{d[\mathrm{HBr}]}{dt} = k_2[\mathrm{Br}][\mathrm{H_2}] - k_{-2}[\mathrm{H}][\mathrm{HBr}] + k_3[\mathrm{H}][\mathrm{Br_2}]. \tag{7.12}$$

Substituting in our expressions for [H] and [Br] (equations (7.11) and (7.9)) gives

$$\frac{d[HBr]}{dt} = \frac{2k_2(k_1/k_4)^{1/2}[Br_2]^{1/2}[H_2]}{1 + (k_{-2}/k_3)[HBr]/[Br_2]}.$$ (7.13)

We see that this agrees with the measured rate law, equation (7.4). In the early stages of the reaction, the concentration of the HBr product is much lower than that of the reactant Br_2, and the second term in the denominator becomes negligible. The rate law then reduces to

$$\frac{d[HBr]}{dt} = 2k_2(k_1/k_4)^{1/2}[Br_2]^{1/2}[H_2]$$ (7.14)

again reproducing the experimental observations. The proposed mechanism therefore fits well with the experimental measurements.

7.2.2 The hydrogen–chlorine reaction

The mechanism for the hydrogen–chlorine reaction is similar to that for the hydrogen–bromine reaction

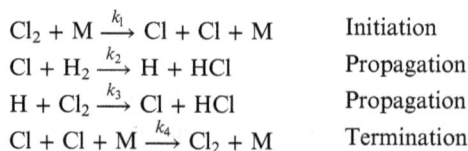

$$Cl_2 + M \xrightarrow{k_1} Cl + Cl + M \qquad \text{Initiation}$$
$$Cl + H_2 \xrightarrow{k_2} H + HCl \qquad \text{Propagation}$$
$$H + Cl_2 \xrightarrow{k_3} Cl + HCl \qquad \text{Propagation}$$
$$Cl + Cl + M \xrightarrow{k_4} Cl_2 + M \qquad \text{Termination}$$

As in the case of the $H_2 + Br_2$ reaction, the reaction may also be initiated photochemically. However, unlike the $H_2 + Br_2$ reaction, both propagation steps are very efficient, and the inhibition step is so slow that we have omitted it from the mechanism. In combination, these factors result in the rate for the $H_2 + Cl_2$ reaction being much higher than that for $H_2 + Br_2$. Chain lengths up to 10^6 are possible, and, coupled with the exothermicity of the reaction, these can lead to a thermal explosion (for more on explosions, see section 7.4).

The $H_2 + Cl_2$ reaction provides a good example of a case for which the steady-state approximation breaks down. Since both propagation steps are very efficient, when the radical concentrations [H] and [Cl] are much lower than the reactant concentrations $[H_2]$ and $[Cl_2]$, reactive collisions between a Cl atom and an H_2 molecule (propagation) are much more likely than terminating collisions of Cl with another Cl atom. This means that the reaction may be very advanced before the steady-state condition—that the rates of initiation and termination are equal—is reached. In practice, the situation is usually simplified somewhat, due to the extreme sensitivity of the reaction to inhibition by contaminants, such as O_2. Molecular oxygen reacts rapidly with H and Cl radicals to form inert radicals, providing a number of alternative termination pathways and increasing the overall rate of termination. If O_2 is present at a concentration of around 1% or greater, the following termination steps dominate to the point where the steady-state approximation becomes valid

$$H + O_2 + M \xrightarrow{k_5} HO_2 + M$$
$$Cl + O_2 + M \xrightarrow{k_6} ClO_2 + M. \tag{7.15}$$

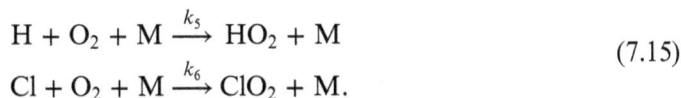

If we replace the termination step (step 4) of the original reaction mechanism with the termination steps above, we can apply the steady-state approximation to obtain expressions for the steady state H and Cl concentrations and therefore (after some algebra, left as an exercise for the keen reader) for the overall rate

$$\frac{1}{2}\frac{d[HCl]}{dt} = \frac{2k_a[H_2][Cl_2]^{1/2}}{[O_2]([H_2] + k_b[Cl_2])} \tag{7.16}$$

with

$$k_a = \frac{k_1 k_2}{k_5} \tag{7.17}$$

and

$$k_b = \frac{k_3 k_6}{k_2 k_5}. \tag{7.18}$$

7.2.3 The hydrogen–iodine reaction

We might expect the $H_2 + I_2$ reaction to have a similar mechanism to the Br_2 and Cl_2 analogues. However, the second step in the mechanism, $I + H_2 \longrightarrow H + HI$, occurs much too slowly at normal temperatures for this mechanism to be viable. Various kinetic mechanisms operate at different temperatures; for example,

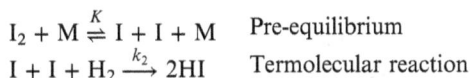

$$I_2 + M \overset{K}{\rightleftharpoons} I + I + M \quad \text{Pre-equilibrium}$$
$$I + I + H_2 \xrightarrow{k_2} 2HI \quad \text{Termolecular reaction}$$

Unlike the complicated rate laws followed by the chlorine and bromine reactions, the mechanism above gives rise to a simple second order rate law (this can be proved as an exercise using the pre-equilibrium treated in section 5.2 as an example)

$$\frac{d[HI]}{dt} = 2k_2 K [I_2][H_2]. \tag{7.19}$$

7.3 Comparison between the hydrogen–halogen reactions

The key difference between the reactions of Cl_2, Br_2 and I_2 with molecular hydrogen lies in the relative endothermicity of the reactions between the halogen atoms and H_2 (step 2 in the chain reaction mechanisms)

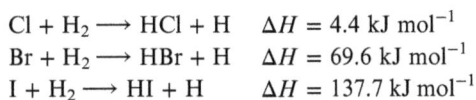

$$Cl + H_2 \longrightarrow HCl + H \quad \Delta H = 4.4 \text{ kJ mol}^{-1}$$
$$Br + H_2 \longrightarrow HBr + H \quad \Delta H = 69.6 \text{ kJ mol}^{-1}$$
$$I + H_2 \longrightarrow HI + H \quad \Delta H = 137.7 \text{ kJ mol}^{-1}$$

The iodine reaction is highly endothermic, and consequently so slow that it does not occur at accessible temperatures, ruling out the chain mechanism observed for the other two halogens. The chlorine reaction is almost thermoneutral and, therefore, fast, and the bromine reaction is relatively endothermic and slow. While the reverse (inhibition) step $H + HX \longrightarrow X + H_2$ is thermodynamically favourable for both Cl and Br, it is unimportant for Cl, due to the very low steady-state concentration of H atoms. In contrast, the inhibition step has a considerable effect on the overall kinetics of the bromine reaction.

7.4 Explosions and branched chain reactions

Explosions without doubt represent one of the most exciting phenomena in chemistry, and often play a significant role in encouraging an early interest in chemistry amongst aspiring scientists. However, for the most part, professional chemists and chemical engineers are rather keen not to experience explosions as they go about their daily business, and it is therefore very important to understand the conditions under which a reaction mixture can become explosive, in order that such conditions can be avoided if at all possible.

An explosion occurs when a chemical reaction accelerates out of control. As the reaction speeds up, gaseous products are formed in larger and larger amounts, and more and more heat is generated. The rapid liberation of heat causes the gases to expand, generating extremely high pressures, and it is this sudden formation of a huge volume of expanded gas that constitutes the explosion. The resulting pressure wave travels at very high speeds, often much faster than the speed of sound, generating a supersonic shock wave or 'bang'.

There are two different mechanisms that may lead to an explosion. These are related to the fact that the overall reaction rate depends on both the magnitude of the rate constant and the amounts of reactants present in the reaction mixture.

If the heat generated in a reaction due to the reaction exothermicity cannot be dissipated sufficiently rapidly, the temperature of the reaction mixture increases. This increases the rate constant and, therefore, the reaction rate, producing more heat and accelerating the reaction rate still further, and so the cycle continues until an explosion results. Such explosions are known as *thermal explosions*, and in principle may occur whenever the rate of heat production by a reaction mixture exceeds the rate of heat loss to the surroundings—often the walls of the reaction vessel. This type of explosion is unfortunately relatively common when a chemist 'scales up' a reaction that has previously been carried out in small quantities, without properly taking into account the laws of thermodynamics.

The second category of explosion arises from chain branching within a chain reaction. Explosions in this category are known as *chain branching explosions*, or sometimes, somewhat misleadingly, *isothermal explosions*. In this case, one or more steps in the reaction mechanism produce two or more chain carriers from a single chain carrier. As the reaction proceeds, the number of chain carriers increases exponentially, leading to a rapid acceleration in the overall reaction rate.

In practice, both mechanisms often occur simultaneously, since any acceleration in the rate of an exothermic reaction due to chain branching will eventually lead to an increase in temperature. It is very important to note, however, that chain branching is not a requirement for an explosion. As an example, the detonation of TNT (2,4,6-trinitrotoluene) is simply the result of an extremely fast chemical decomposition that generates huge quantities of gas.

The reaction $2H_{2(g)} + O_{2(g)} \longrightarrow 2H_2O_{(g)}$ provides an example of a reaction in which both mechanisms are important, and we will consider its mechanism in some detail.

7.4.1 The reaction $2H_{2(g)} + O_{2(g)} \longrightarrow 2H_2O_{(g)}$

Though the reaction between molecular oxygen and molecular hydrogen may appear to be very simple, the mechanism is in fact extremely complex, and is still not fully understood. The process is known to be a branched chain reaction, involving carriers such as H, O, and OH. A simplified version of the thermal mechanism is given below

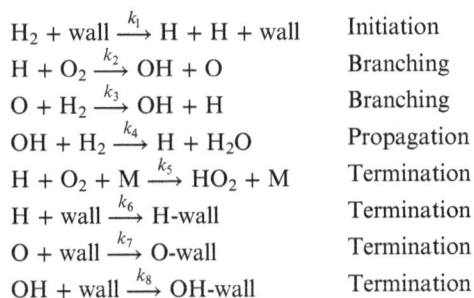

$$
\begin{array}{ll}
H_2 + \text{wall} \xrightarrow{k_1} H + H + \text{wall} & \text{Initiation} \\
H + O_2 \xrightarrow{k_2} OH + O & \text{Branching} \\
O + H_2 \xrightarrow{k_3} OH + H & \text{Branching} \\
OH + H_2 \xrightarrow{k_4} H + H_2O & \text{Propagation} \\
H + O_2 + M \xrightarrow{k_5} HO_2 + M & \text{Termination} \\
H + \text{wall} \xrightarrow{k_6} \text{H-wall} & \text{Termination} \\
O + \text{wall} \xrightarrow{k_7} \text{O-wall} & \text{Termination} \\
OH + \text{wall} \xrightarrow{k_8} \text{OH-wall} & \text{Termination}
\end{array}
$$

Reaction steps 2 and 3 in the mechanism both produce two radical chain carriers for each chain carrier consumed. The two steps combine to give an overall branching coefficient of three, corresponding to the hypothetical reaction $H + O_2 + H_2 \longrightarrow H + OH + OH$. Even so, the mixture is not explosive under all conditions, mainly because steps 1, 2, and 3 are endothermic ($\Delta H_1 = 427$ kJ mol^{-1}, $\Delta H_2 = 71$ kJ mol^{-1}, $\Delta H_3 = 17$ kJ mol^{-1}) and are therefore slow at low temperatures. The efficiency of the branching steps increases with increasing temperature, and as a result, the reaction displays a complex dependence on temperature and pressure, as shown in figure 7.1. We will explore this dependence in the following.

Figure 7.1 shows the regions of temperature and pressure over which the reaction is explosive (shaded regions) or proceeds steadily without exploding (unshaded regions). Consider the vertical dotted line on the diagram. Travelling vertically from the bottom to the top of the line corresponds to gradually increasing the pressure at a fixed temperature of 800 K. We see that at low pressures the reaction proceeds steadily, then reaches a range of pressures over which it becomes explosive, before returning to steady reaction at higher pressures, and finally becoming explosive again at very high pressures. The regions are demarcated by a number of *explosion limits*, also marked on the diagram. It is relatively straightforward to explain the presence of the various explosion limits by reference to the reaction mechanism.

1. At very low pressures the mean free path in the gas is large, and chain carriers can reach the walls and combine. Collisions with walls are more

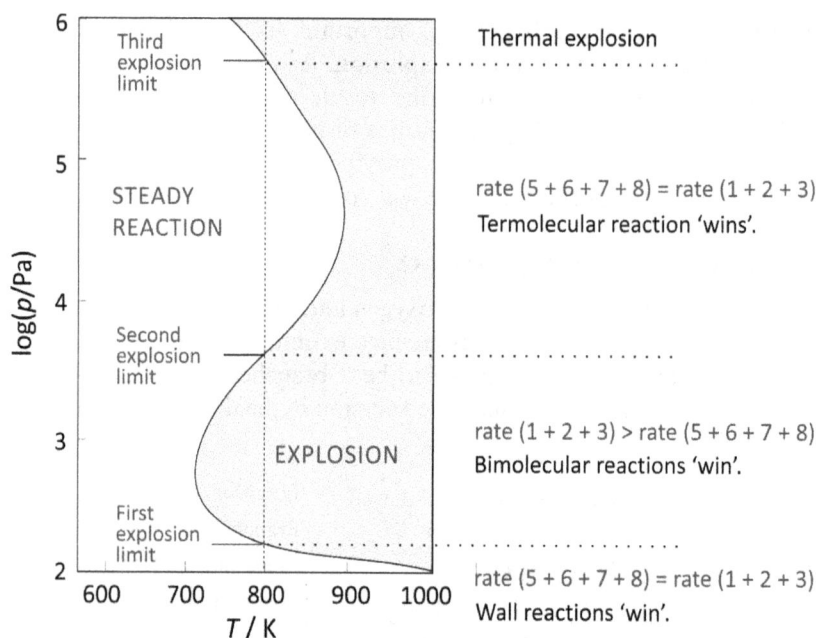

Figure 7.1. Explosion diagram for the reaction between H_2 and O_2.

likely than collisions with other gas-phase molecules, so that overall the rate of steps 5, 6, 7, and 8 balances that of steps 1, 2, and 3, i.e. termination balances initiation, and steady reaction occurs.

2. At higher pressures, the chain carriers react before reaching the walls, and the gas-phase branching ratios become too high for wall termination to control. This is primarily due to the fact that the rates of the bimolecular gas-phase reactions are proportional to (pressure)^2, while the rate of wall reactions increases in direct proportion to the pressure. As the pressure increases, the rate of reactions 1, 2, and 3 exceeds that of reactions 5, 6, 7, and 8, and the mixture becomes explosive once the pressure reaches the first explosion limit.

3. At even higher pressures, three body collisions start to become important. Termination step 5, whose rate is proportional to (pressure)^3, can now match steps 1 to 3 in efficiency. The collective rate of reactions 5, 6, 7, and 8 now outweighs the rate of steps 1, 2, and 3, leading to steady reaction at pressures above the second explosion limit.

4. If the pressure is increased still further, the reaction rate increases so much, and such a large amount of heat is generated, that a thermal explosion results above the third explosion limit.

We can also see from figure 7.1 that if we change the temperature, the explosion limits occur at different pressures. Other factors, such as the shape and size of the reaction vessel, and the presence of an inert gas, can also change the explosion limits.

These effects are fairly straightforward to understand if we consider the likely effect of each variable on the relative rates of initiation, propagation/branching, and termination steps.

1. *Temperature*

   Increasing the temperature increases the efficiency both of endothermic reaction steps and steps for which there is an activation barrier. This applies to steps 1–4 in the above mechanism. The termination steps are less sensitive to temperature, and may even be slowed down since they tend to be exothermic. As a result, the first explosion limit is lowered, as the rates of steps 2 and 3 outpace those of steps 6–8 more readily. The second explosion limit is increased because at higher temperatures a higher pressure is needed in order for the termolecular step 5 to become important. The third limit is decreased, since at higher temperatures more heat is produced, and also the heat that is produced is harder to lose from the system.

2. *Surface-to-volume ratio*

   The shape and size of the reaction vessel can have a considerable effect on the explosion limits. Increasing the surface-to-volume ratio favours processes that involve collisions with the vessel walls over gas phase processes, which in this case means the initiation step 1 and termination steps 6, 7, and 8. The high efficiency of the branching steps means that step 1 is unimportant in determining the overall rate, and therefore the increased efficiency of the termination steps increases the pressure at which the first explosion limit is reached. The second limit has no dependence on the vessel walls and is unchanged. The third limit increases because it becomes easier to lose heat from the system due to the greater number of collisions with the walls.

3. *Total pressure*

   Adding an inert gas to the mixture decreases the mean free path of the gas molecules, and disfavours collisions with the walls. This lowers the first explosion limit, since the termination steps 6, 7, and 8 become less efficient. The second limit is also decreased because the inert gas can act as the third body M, increasing the rate of step 5, a termination process. The third limit is lowered due to the reduced heat transfer from the gas to the vessel walls.

We have seen that a number of different variables come into play when determining whether or not a reaction mixture is likely to become explosive. These should all be considered carefully when working with any potentially explosive reactions in order to minimise any safety risks.

7.5 Concluding remarks

Over the course of this short introduction to chemical kinetics, we have explored in some detail the concepts of elementary and complex reactions, defined the rate of reaction, and considered the various factors on which the rate depends. We have shown that the reaction rate may be related to reactant concentrations through the relevant rate law, and to temperature through the Arrhenius equation. We have looked at a number of experimental techniques for measuring reaction rates, and a variety of analysis methods that allow us to determine the rate law from experimental data. Finally, we have explored the link between the rate law and the reaction mechanism, and have demonstrated the mechanistic insight that can be gained from kinetics measurements by means of a variety of examples, including unimolecular reactions, third-order reactions, enzyme reactions, chain reactions, and explosive reactions.

In closing, it is worth noting some of the aspects of reaction kinetics that we have *not* covered. Simple collision theory, covered in chapter 1, is the simplest theoretical model developed to interpret and predict chemical reaction rates. It treats the reactants as 'hard spheres', and does not include the effects of any long-range interactions—for example van der Waals or ionic interactions—or any internal degrees of freedom of the reactants. A number of much more sophisticated models are available, some of which we have alluded to at appropriate points in the text. These include transition-state theory, RRK and RRKM theory, and capture theories, all of which are discussed in more advanced texts on reaction kinetics (see section 7.6).

In focusing mainly on reactions occurring in the gas phase, molecular interactions other than those driving the reaction itself can be ignored to a very good approximation. However, for reactions in solution, particularly those involving ionic reactants, the effects of the solvent on the reaction can be significant. Reactants must diffuse together through solution before they can react, and newly-formed products find themselves confined within a 'solvent cage'. Factors such as solvent viscosity and polarity can have a dramatic effect on the reaction rate, and in some cases the solvent may play an important role in the reaction mechanism.

Reactions occurring at surfaces also deserve a mention. As well as forming the basis of many catalytic processes, this category of reactions also encompasses the field of electrochemistry. Unsurprisingly, the reaction rate can depend critically on the properties of the surface, as well as on the properties of the surrounding gas, liquid, or solution phase.

While these scenarios certainly present additional challenges, theories and models developed to describe reaction rates within these more complex environments all build on the fundamental framework outlined in the present text. Having mastered the material in this book, the reader should therefore be well equipped to explore more advanced treatments of kinetics. A few suggestions for further reading are listed in section 7.6.

7.6 Further reading

Atkins P W and de Paula J 2014 *Atkins' Physical Chemistry* 10th edn (Oxford University Press)
Pilling M J and Seakins P W 1996 *Reaction Kinetics* 2nd edn (Oxford University Press)
Laidler K J 1997 *Chemical Kinetics* 3rd edn (Pearson)
Mortimer M and Taylor P G 2004 *Chemical Kinetics and Mechanism* (RSC Press)
Cox B G 1994 *Modern Liquid Phase Kinetics* (Oxford University Press)
Houston P L 2006 *Chemical Kinetics and Dynamics* (Dover Publications)
Brouard M 1998 *Reaction Dynamics* (Oxford University Press)
Albery J 1975 *Electrode Kinetics* (Oxford University Press)

www.ingramcontent.com/pod-product-compliance
Lightning Source LLC
Chambersburg PA
CBHW082111210326
41599CB00033B/6671